In Our Own Image

Humanity's Quest for Divinity via Technology

by
Debashis Chowdhury

Bloomington, IN  Milton Keynes, UK

authorHOUSE

AuthorHouse™
1663 Liberty Drive, Suite 200
Bloomington, IN 47403
www.authorhouse.com
Phone: 1-800-839-8640

AuthorHouse™ UK Ltd.
500 Avebury Boulevard
Central Milton Keynes, MK9 2BE
www.authorhouse.co.uk
Phone: 08001974150

First published by AuthorHouse 2/21/2007

ISBN: 978-1-4259-4279-3 (sc)
ISBN: 978-1-4259-4280-9 (hc)

Library of Congress Control Number: 2006905388

Printed in the United States of America
Bloomington, Indiana

This book is printed on acid-free paper.

Preface
In Our Own Image

Humanity's Quest for Divinity via Technology

It is human nature to maximize the utility of the knowledge that we have, and minimize the exposure to the knowledge that we lack. As a result, we tend to build our knowledge in a contiguous manner – adding a bit here, patching a hole there – and trusting that somehow, someway it all makes sense. Even the scientific method is designed to build on what we already know. What suffers as a result is a view of the all encompassing whole – how physics connects to the metaphysics, how our mind connects to our body, and how our microcosm (the human body) connects to the macrocosm (the universe).

In this book we will take a different approach. The principal goal of this book is to connect us - as humans, individually and collectively, to the expansive reality that we see around us. In seeking to transcend our current condition, we need to know what parts of our existence are uniquely human – and thereby held in common with the 6+ billion others with whom we share this planet. We also seek to understand those fundamental constructs which can be held relatively constant through various spheres of human existence; as well as other equally fundamental patterns whose persistent 'rhythm' permeates our reality – all the way from the sub-atomic to the universal.

We will start off Chapter 1 with the discussion on the patterns that make us uniquely human. The concept of the Mahashrama cycle, derived from ancient Indian philosophy, is a key 'Image' that we will carry with us throughout this book. The original Ashrama system (upon which Mahashrama is based) divides a person's existence into four segments, Brahmacharya (student-hood), Grihastha (householder, economic engine), Vanaprastha (service leadership) and Sanyas (spirituality). Although we retain the same four segments from the original Ashrama, the Mahashrama

formulation adds in the competitive aspects of the first two segments, thus bringing it up to date with the highly competitive reality we see around us today.

Drawing from a different source of inspiration, the Bible talks about God having created man 'In His own image'. If we believe in our 'God' as creator – is it possible that we can find the 'image' and create a 'God-like' progeny civilization of our own? If we do not believe in a creator, is it possible for us to take the core essence of what it means to be human, and use it to seed a civilization that has the potential to grow up and set it's own standard of 'Divinity'? In the second scenario, could we continue scaling up until we have infused the entire universe with Divine purpose? In either scenario, it is important to find the potent 'Image' that has the power of a near-infinite magnification. In Chapter 1 we will also discover that hidden in the workings of Mahashrama is the ability to accumulate enough knowledge and wisdom (Cultural DNA) that it can prepare us to jump to the next level of existence at the appropriate time.

In Chapter 2, we apply the concepts of Mahashrama to the engine of modern economic productivity – the corporation. Not only are corporations the economic engine, but they are also driving the technology that allows us to have a wider reach amongst all humans. This is happening primarily through the internet – which has turned the traditional value delivery system (generating and charging for services) on its head. The result is that human beings from all walks of life can now connect, and get their needs (information and otherwise) potentially satisfied in a highly individualized manner. It goes without saying - the 'reach' of the internet is critical to the scalability of our human 'Image' towards our next level of existence. To avoid splintering the human race we must strive to get all of humanity find a common identity and purpose - as we get ready to collectively frame our civilizational future.

We look at the principles of Governance in Chapter 3, including the major shift in paradigm that we find happening around us globally. At this juncture in history we find that the majority of humanity has not yet made the transition over to democracy – and

even where it has, we are sometimes subject to the 'tyranny of the masses'. We look at some of the great cultural conflicts that have arisen between the Theocracies and the Democracies of the world, and introduce the principle of Mutually Assured Preservation (MAP), to resolve conflicts arising from divergent worldviews. Using the Mahashrama model, we notice an unusual affinity between the opposite segments – leading to a clear polarization of the world along the two worldviews of Aristocracy-Theocracy and Individualized Democracy. Here we develop the concept of Individualized Democracy, where the constituents are individually kept abreast of all the issues that are of impact to them – and are in turn able to directly interact in framing the issues of the day. New technology will definitely be needed to support this kind of inclusive governance. To illustrate how individuals and groups can have their say, even if they do not directly control huge voting power, we examine the concept of the 'diversity vote' in the context of the presidential elections of the USA.

For a vibrant society that has internalized the Mahashrama cycle, it is important to know what the key needs are of the population, who are at various stages of Mahashrama development. In Chapter 4 we start off with the 'Hierarchy of Needs' developed by Maslow, and develop it to cover all the Mahashrama segments. We introduce 'Yogic Empathy' as the key skill by which the Service-leadership segment within Mahashrama, namely Vanaprastha, can relate to the needs of the entire population, including the spiritually advanced 'sanyasi'. Here we also look at the concept of accumulating Cultural DNA as the culmination of human learning, and the process of creation of the Human Cultural Genome. This is the internally consistent rendition of what it means to be human, and cuts across barriers of race, gender and ethnicity. Needless to say, technology is a big part of it, and if we do our job right - the Human Cultural Genome can become the 'functional DNA' of the next level of our existence.

The 'glue' that holds the population together at the societal level is made up of the service leaders from the Vanaprastha segment. In Chapter 5 we look at how the individuals within the Vanaprastha segment organize to form the Vanaprastha Network – which can then develop an individual relationship with every human being

in a given population. We compare the Vanaprastha Network to the workings of the human brain, and the collective action of billions of 'neurons' or nerve cells that can serve as a model for developing our own civilizational 'brain'. We also look at some of the technology trends that will enable the Vanaprastha Network to respond to the smallest individual needs, and yet are sufficiently scalable to meet the challenges of humanity in a global scale.

In Chapter 6 we look at the unique challenges and opportunities in the final segment of Mahashrama – Sanyas. Sanyas deals with spirituality, and its utility is the least understood today, given our current materialistic cultural mindset. Sanyas grapples with our ability to 'transcend' our current situation in favor of a vastly more attractive reality that lies beyond our normal day-to-day experience. In this chapter we take on the 'hallowed ground' of religious diversity, and demonstrate that our current system of beliefs are but human responses to alternate spiritual realities – as viewed through the eyes of Mahashrama. We introduce the concept of Yogic Universality by which all Yogic capable entities and intelligences can join together to form a 'whole' that is immeasurably bigger than the sum of the parts.

Chapter 7 deals with the technology and philosophy behind 'preservation' and 'reenactment' – which together constitute a form of immortality. We recognize that the technology to store every relevant multimedia fact in our lives is already coming on line – and that in the future we can access 'out-of-body' networks to get a unique vantage point for our lives. We discuss the skills of Virtual Presence, and discover how our children are already getting good at this skill-set through the use of virtual reality and computer games. To conclude the chapter, we look at a series of 'gasket' interfaces that would allow the human mind to interface with machines at progressively higher levels of abstraction.

As we build towards our future, it is important to understand some of the persistent patterns that surround us, and their possible implications on a future transcendental level of existence – which we refer to as the Supra (or Supra Human Entity). In Chapter 8, we examine eight sets of patterns from around the Mahashrama cycle, ranging from the microscopic to the astronomical. Based on this progression, we discover that the scaling factor from human to Supra that we have just inferred – can be repeated four more times before we reach the magnitude of the visible universe. Moving up the ladder, from atoms to cells to humans, we will see that at each stage, uniformity decreases and differentiation increases. Using this progression, we will estimate how much internal consistency we will need in the DNA of our Supra entity for it to be able to function effectively.

Having discovered the patterns surrounding our technological existence, in Chapter 9 we look at the design attributes that we need to apply to create the Machine Supra Cell (MSC). Drawing from cellular biology, the MSC is the machine counterpart to us humans, and will be the basic unit of intelligence that will subsequently form the Supra entity. The reason for keeping approximate parity between humans and MSCs is very important – and determines if we can continue on as symbiotic entities, or if machines would one day take over our entire civilizational trajectory. There are 18 design attributes listed, and might look like a dense technical compilation for some readers – who could choose to move directly to the chapter conclusion.

In Chapter 10 we correlate several of the patterns discussed in Chapter 8, and show how they connect us humans from the microscopic to the astronomical. We dig deeper into the Solar Local Ratio of 108 – and discover a deep significance within its factorization sequence. Using this sequence, we can surmise the nature of the 'Tree of Universality' that connects us from the mundane to the transcendental. Drawing inspiration from the 'Inverted Tree' described in the Bhagavat Gita (ancient Indian philosophical discourse) – we look at the significance of the various levels, and notice a very remarkable correlation between the two trees. Some of the mathematical analysis and analogies are carried over to Appendix 2.

The concluding chapter (Chapter 11) is where we tie together the relevance of the key arguments made in this book. Moving forward, we note the critical importance of getting most (if not all) of humanity on the same page as to the nature and purpose of our collective human existence. We refer to this process of

generating a level of consensus regarding the 'Identity' and 'Purpose' of our combined human existence as the 'planting' of the Tree of Universality. After observing how truly complementary our two sets of progeny, human and machine, can be – we note that together they form a highly responsive and scalable fabric that can reach for the stars. We also look at some of the key short to medium term tasks (next 100 years) that humanity must undertake to ensure both our survival and prosperity. These are important steps as we try and secure our long term future through transcendence into a higher level of existence. Lastly, we discuss the different types of Intelligences that have been attributed to humans – the Intelligence Quotient (IQ), the Emotional Quotient (EQ) and the Spiritual Quotient (SQ). As we prepare to move on towards a level of 'Divinity', the ability to organize our civilization effectively and move us forward collectively will become a new measure of intelligence for our Civilization – which we will refer to as the Philosophical Quotient (PQ).

Apart from creating a framework that integrates the various aspects of human endeavor – what other objective should we have from our reading of this book? It is my deeply felt aspiration that the reader will also come away with a feel for how humanity is poised for success or failure as a race and as a civilization. The timeframes we use in this book may look quite long (short term is 25 years, medium term 100 years), but they are necessary to paint the picture of the civilizational cross-roads that we find ourselves in. The choices are stark and unambiguous. The first choice is to figure out a common ground for humanity as a civilization and as a race, and work towards a commonly subscribed vision of the future that moves us towards 'Divinity'. The other option is to continue with the religious, ethnic and economic conflicts that sap so much energy from our civilizational brain that we end up missing the transformational opportunities altogether. Technology is a double edged sword – with the right civilizational preparation it can take us to heights that we can only dream of today. Without adequate preparation, technology will take on a character of its own, which either emulates our conflicting ways – or more likely sits in judgment of our continuing bad habits. You see, our machine progeny will be essentially immortal, and they will get the long view! Our machine progeny will also not be confined to an

atmosphere and to a biological support system, and will be free to roam the solar system. It will take effort to integrate humanity into all aspects of a future machine civilization – and the cost-benefits of human inclusiveness will need to be carefully weighed. If we demonstrate an inability to transcend our genetic programming, 'they' will be the ones who determine if there is a continuing role for us – or whether we should be quarantined and left to our habits on a lonely planet called Earth.

In the end, there is really one logical conclusion. The Mahashrama Cycle and the Tree of Universality provide us with a framework using which mankind can 'transcend' our current existence. They provide an umbrella under which we can integrate the diversity that we have as a civilization, and move forward purposefully as a single civilization. Planning for a 'symbiotic' relationship between our human and machine progeny gives humanity the 'jump' we need to not only survive, but also to prosper - as a race, as a civilization, and as a transcendental presence here in this universe!

For updates on our progress as we undertake this remarkable quest, as well as a personal tool for you to assess your own Mahashrama priorities, please check the following URL: www.iooi.org

Table of Contents

Glossary of Terms:

Ashrama – Original Indian Philosophy covering the stages of human existence.

Mahashrama – Extended (or Greater) Ashrama, enhanced to include competitive forces. The four segments of Ashrama and Mahashrama are the same, and listed below:

Brahmacharya – The first segment of Mahashrama. Emphasis is on learning and proficiency in the skill set required for competitive self sufficiency. The 'Individuality' of the human being is also developed at this stage. The person going through the Brahmacharya stage is referred to as the 'brahmachari'.

Grihastha – The second segment of Mahashrama. Emphasis is on discharging our Genetic Imperative (procreation) and Economic Imperative (providing for family, future activities). The person going through the Grihastha segment is the 'grihasthi'.

Vanaprastha – The third segment of Mahashrama. Emphasis is on Community and Service leadership, with the goal of growing spirituality to be able to relate to causes that are important to all of humanity. The person going through Vanaprastha is the 'vanaprasthi'.

Sanyas – The final Segment of Mahashrama. Emphasis is on spirituality with the goal of 'transcending' our human condition, and codifying the 'essence' or 'spirit' of what it means to be human for posterity. The person going through Sanyas is the 'sanyasi'.

DNA – Deoxyribonucleic Acid – the material out of which the genetic sequences (genes) within our chromosomes is constructed. In general - our inherited genetic content, or Genome. Sometimes referred to the 'Nature' component of who we are as humans.

Cultural DNA – The sum total of learnings (from experience) for the individual human being. Some effort is required to separate the experiences that really have meaning – and those that are merely incidental. Sometimes referred to as the 'Nurture' component of who we are as humans.

Cultural DNA (Ethnic) – Cultural (or experiential) learnings that define the commonly held experiences, learnings and beliefs of an Ethnic group or Community. Quite a bit of effort is usually required to crystallize out the parts that are really relevant to, and representative of the character of the ethnic group or community. .

Cultural DNA (Human) – Sum total of all human experiential learnings. Over time, the Cultural DNA will need to become condensed down to the most concise and internally consistent thread(s) – eventually giving rise to the DNA for the next level of our existence.

Supra Human Entity – Projected next level of human existence with 'Divine' capabilities. Expected complexity level is about 10^{15} human intelligences. Also referred to as the 'Supra'.

Machine Supra Cell (MSC or Supracell) – Unit of Machine intelligence with about the same degree of sophistication as a human being. The MSC then becomes the cellular unit (along with humans) for a future Supra Human Entity.

Yoga – Spiritual Union. From Indian philosophy, this is the mechanism by which two dissimilar quantities (e.g. mind and body, material and spiritual) can exist and perform as a single entity.

Yogic Union – the combination of two entities that is greater than the sum of the parts, and which allows the composite entity to behave as a single entity.

Yogic Empathy – Fellow-feeling, sharing and openness that comes from Yogic union.

Yogic Unification – The process of applying Yogic Union to various constituents of a population such that they all feel part of the overall organism or entity.

Yogic Universality – The logical end point of Yogic Unification, when all of the intelligence within this Universe is part of a single Yogic organism.

Mutually Assured Preservation (MAP) – The principle of conflict resolution that uses Yogic Empathy, and where the continuing existence of all parties in the conflict remain preserved.

Varna – Ancient Indian Caste system, by which the social responsibilities of a family was determined by birth (e.g. Brahmins were priests, Ksatriyas were warriors and rulers).

Ananda – Spiritual Joy. A state of happiness one can aspire to with adequate spiritual preparation.

Serial Endosymbiosis Theory (SET) – The theory of how the first Eukaryotic cells (with a clearly defined cell nucleus) formed. These cells became the building blocks from which all of modern day plants and animals are constituted. The theory proposes that several cellular biological features evolved independently, and were symbiotically fused together to form a superior organism (i.e. the first Eukaryotic cells).

Vanaprastha Network – The organization formed when the vanaprasthi, who seek the betterment of the universal human condition, organize to become the 'brain' and nervous system of a future human society.

Dvaita (Dualism) – the belief that God and Humans are forever different and separate. In this belief, we can come close to God, but cannot become 'one' with our God. Sometimes also applied to the perceived permanent differences between our material and spiritual existence.

Advaita (non-Dualism) – the belief that we humans are part of the core spirituality of the Universe (or God), and differ only in the degree of 'Godliness'. With spiritual development, one can hope for total spiritual union, or becoming dissolved in this sea of spirituality (Moksha or Nirvana).

Cultural Archeology: Future field of endeavor by which incomplete Cultural Genomes (individual, Ethnic or Civilizational) can be reconstructed and recreated to lifelike realism.

Prime Ascension – The concept that humanity can achieve a level of Divinity (Supra Human Existence) without the help of other (non-human) Intelligences. If we do get external help, we can still have Ascension, but it will be assisted.

Solar Local Progression Ratio (108) – The ratio between the Earth, the Sun and the Earth-Sun distance, that hints at a level of spiritual organization.

Self Power Factorial (SPF) series – $1^1 \times 2^2 \times 3^3 \ldots N^N$. Derived from the Solar Local Progression Ratio, 108, which equals $1^1 \times 2^2 \times 3^3$.

Self Power (SP) function - N^N – This is a derivative of the SPF, and useful for the calculation of certain combinations between Yogic Entities.

Tree of Universality – A mathematical model of the structure of reality that connects the mundane (Level 3) to the Transcendental (Level 0). Derived from the Solar Local Progression Ratio.

Veil of Abstraction – The interface between Level 1 and Level 0 in the Tree of Universality that serves to abstract (or isolate) the reality below from the spirituality above.

Grihastha *Vanaprastha*
Brahmacharya *Sanyas*

Acknowledgements

I truly appreciate the support of my family, who spared me
the time and effort to complete this book, and offered me the
encouragement to persevere in the long road to completion. This
work would not have been possible without editorial help from my
wife, Sarbari. In addition, if you find the concepts developed here
as relevant and pertinent to our human condition – this is only
because she helped me refine the concepts through her unique
style of analytical inquiry. Her inexhaustible curiosity about other
countries and societies has enabled our family to get exposure to
a wide variety of cultures. Indeed, it is one of those trips (to Peru
in 1998) that the idea came to me that there were some long-term
ideas about civilizational success and failure that needed to be
developed. You see the results of this multi-year project organized
and presented in this book.

I also want to thank our daughter, Shilpika, for her help with the
art-work. My introduction to Indian spirituality came at an early
age, and for this I thank my mother. She even arranged for me
to have a Guru or 'Spiritual Adviser' as the traditional Indian path
to spirituality. As an interesting side note, the 'Sampradaya' or
'spiritual lineage' that my grandparents drew their inspiration from,
was founded by Sri Nimbarka – whose spiritual philosophy is
unique in that it unites the apparent contradictions between 'dvaita'
and 'advaita'. These concepts are discussed further in Chapter
6. My father, by contrast, has always been a pragmatist – and to
him I owe the analytical desire to 'engineer' something better than
what we have today.

I would like to express my gratitude to the following reviewers for
taking time out from their busy lives to read my early drafts and
give me feedback.

Lil Mohan
Partha Dasgupta
Tushar K. Ray, Ph.D
Utpalendu Chowdhury

There are numerous others who have been a source of inspiration and support, and I cannot even begin to name them all. Finally, I would like to thank my employer, Intel Corporation, for giving me the opportunity to experience a range of challenging vocations, including Chip Design, Product Marketing, Researching new computer usages, Strategic Planning and even Venture Capital Investing (as part of Intel Capital).

Grihastha Vanaprastha
Brahmacharya Sanyas

Dedicated to:

My Parents: Umapada and Surama Chowdhury

Chapter 1
The Mahashrama Cycle

"Intelligence is that faculty of mind, by which order is perceived in a situation previously considered disordered."
- Haneef A. Fatmi, founder
Cybernetics Society (London)

"Men often become what they believe themselves to be. If I believe I cannot do something, it makes me incapable of doing it. But when I believe I can, then I acquire the ability to do it even if I didn't have it in the beginning."
- Mahatma Gandhi

Intelligence is indeed all about recognizing the patterns that we see around us, as well as our ability to utilize them as we go about building the best future for ourselves, our children, and the people around us. This book is all about a pattern search, as we try and zero in on some of the larger scale patterns that permeate our life. The patterns we will cover are intended to connect our human existence to the reality around us – all the way from the microscopic to the astronomical. Yet, one of the truly remarkable patterns that we will discuss in this book is uniquely human, and the foundation for any human institutions that we build around us. This is the Mahashrama Cycle – the focus of discussion for our current chapter.

Yet, just recognizing patterns is not enough. Intelligence is also the ability to transform, not only our environment but also ourselves, to become what it is that we believe we have the capability to become. The scale of the Universe is indeed large. We, as humans (the endlessly curious and ambitious entities that we are) will naturally go about determining an inspirational trajectory for our individual existence, as well as for our collective human civilization. As noted by Mahatma Gandhi in his quote above, the power of belief is indeed transformative. Our vision of success will largely determine how deep an imprint we leave upon this Universe with our existence.

Interestingly enough, the human potential is maximized only if we see our future as a confluence of human and machine capabilities. No longer will it be an 'us' and 'them' between our human and our technological offspring – but a synergistic, yogic union between the two, rooted in our common human cultural identity. So, how would we go about building a cultural genetic identity that takes the best of our diverse existence and turns them into building blocks of inspiration for future generations? The answer, here again, is through the use of the Mahashrama Cycle – the fundamental concept around which we will organize everything that is human. We will see that the same patterns of organization hold - from individuals to nations and civilizations. We can extrapolate the Mahashrama pattern well into the future to when our wise and technologically sophisticated human race reaches a level of existence that we would today associate only with Divinity! Immortality, Omnipresence, Omniscience and Omnipotence – yes, it all becomes possible if we start off with the highly scalable 'Image' that is embodied within the Mahashrama Cycle.

The Nature of the Human Animal

Humans are strange animals. For an idea we will break all laws of nature – including the ones that are responsible for our very existence and survival. To all intents and purposes, Homo Sapiens is the product of Evolution – yet our survival seems no longer dependent on the laws of Evolution – either as a species or as individuals. Nature, as expressed in the Biosphere within which we find ourselves, is very efficient and wastes little. We humans often pride ourselves in how much more than our basic necessities we can consume, and think nothing of living outside our budget of 'renewable resources'. The basic tenets of human civilization and medical technology seem to have made obsolete the Natural Selection process that underlies Evolution – 'Survival of the Fittest'. We humans care for our 'unfit' and nurse them back to health. By most measure of 'fitness' (e.g. economic prosperity) – the old measure of 'success' (i.e. the number of children we leave behind). . . today holds a negative correlation!

Chapter 1

We have replaced the Laws of Nature with the Laws of Man. Even these we appear to flaunt on a regular basis. A young man walks up to his father and says, "Dad, working with you has been great, I now know all the rules by which I can be successful in life. I must now go off and make a life for myself."

To which the father answers, "You may know all the rules, but I know the exceptions."

The implication, of course, is that for every rule there exists, there is a way to get around it – legally or illegally. Sometimes one gets caught, and pays the penalty. Most times, society just looks upon such individuals as 'enterprising' and considers them to be successful role models. This behavior will often transcend from individuals to human institutions like Corporations and even Nations. Often times, it seems much more energy goes into figuring out how to beat the 'System' than goes into the very design and upkeep of the system itself.

Yet, in spite of all our human weaknesses, we seem to be well into an age of prosperity based on market economics. Who can deny that the general level of education and economic well being has gone up significantly in the last century throughout most of this world? Or that nations and ethnic groups are fighting and killing each other at a level that as a percentage of the total population is likely at the lowest level we have seen in all of history! Is this the lull before the storm, or a general trend that we can expect to continue well into the next Millennium?

Overall, there is reason for optimism that we have the beginnings of what it takes to build a great future - but the journey will be fraught with tremendous perils. If we can anticipate some of these perils, we can plan ahead and prepare for them. In our journey we will use the Mahashrama Cycle as the philosophical North Star – the one invariant using which we can locate the 'Human' elements that are both our strengths and our weaknesses. Without this instrument, we might forever find ourselves oscillating between the need for prosperity and the need to secure ourselves from those that do not share the prosperity. Isolating ourselves will not work any more! The world is small now and getting smaller in such

a way that problems (and opportunities) are going from local, to regional, to global - very quickly. We are all in this journey together and for the long haul!

The Starting Point

Before we begin to chart our journey, let us reflect on how fortunate we are as a species to be able to undertake this fabulous expedition. The achievements of the human race to date, tremendous as it is, is extremely tiny compared to what we have yet to achieve. Even then, our starting point is in itself quite amazing.

Humans are blessed with a biological miracle as we begin this journey – that we start as a single species throughout all the continents, and there are no other living creatures that we know of that even comes close to our thinking and communicating capability. It is truly remarkable what evolution has gifted us! And it will become even more important as we undertake this journey under very close confines, where the smallest of differences might seem like great divides worth fighting over. Without this small detail of evolutionary pre-history we could well have been chimpanzees and gorillas trying to figure out if this ship is big enough for the both of us!

To complement this biological unity, we also have a tremendous amount of cultural diversity that we will need to carry and preserve on this trip. To a large extent the very way we think and live our lives is dependent on the cumulative experiences of our lifetime to-date. The uniqueness of our upbringing and the effect it has on us can be referred to as the 'Nurture' or cultural part of our upbringing. The contrast, of course, is with 'Nature', our genetic programming or DNA (deoxyribonucleic acid, the building blocks for the genes within our cells) – which is surprisingly uniform. The DNA sequences have been shown to be significantly less than 1% different between any two humans, irrespective of race! The 'Nurture' or Cultural DNA varies widely, but with the popularity of 'Western Culture' throughout the world today, several cultural

viewpoints will likely lose their last adherents over the next half century. In this book we will draw upon several philosophical underpinnings from ancient India as an example of a viewpoint that we must not allow to be lost forever.

Business as Usual

As we have noted, the biggest positive in the current environment, arguably, is the Market Economy that more and more countries are adopting. Following right in the heels of the Market Economy (and maybe a large contributing factor) is the opening up of communications to large sections of the world that remained 'closed' earlier and a continual rising of the level of education - so that people all over the world can now take advantage of this communication.

Why does the Market Economy work so well? This is perhaps the only truly competitive field of endeavor left for humans. The laws of Evolution have worked very well, but have really ceased to be a factor in the choice of which individuals amongst us survives and creates viable offspring in modern society. However, the same laws of 'Survival of the Fittest' now applies to the various products and services made available in the Market Economy. Whole companies and even industries will often cease to exist if they cannot compete – leading to a test of evolutionary 'fitness' that is truly merciless. Also, this could be the only area left where 'Greed is Good' (within limits) and one could unashamedly strut their own prowess at providing the very best products or services to the customer. On a relative scale, the amount of energy devoted into activities like sales and advertising (winning the customer) is just as extravagant as the plumage of the male peacock getting ready for the mating dance!

So why can't the evolutionary forces at work in our Market Economy lead us to both 'Security' and 'Prosperity' in the upcoming millennium? To understand this, let us introduce one of the central themes that we will use throughout this book – the concept of Ashrama or 'Stations of Life' drawn from Ancient Indian philosophy, and enhanced to include today's competitive forces, and economic reality. Let us refer to this expanded

cycle as Mahashrama – meaning Maha (greater) Ashrama. Enhanced in this way, the Mahashrama Cycle can be applied to not only individual humans, but also to human institutions like Corporations, Nations and even entire Civilizations! The stakes are indeed high, and we will see that the 'security' we all desire only comes from the empathic and cooperative side of our nature – which is seriously under-developed in our current cultural mindset.

The Mahashrama Cycle

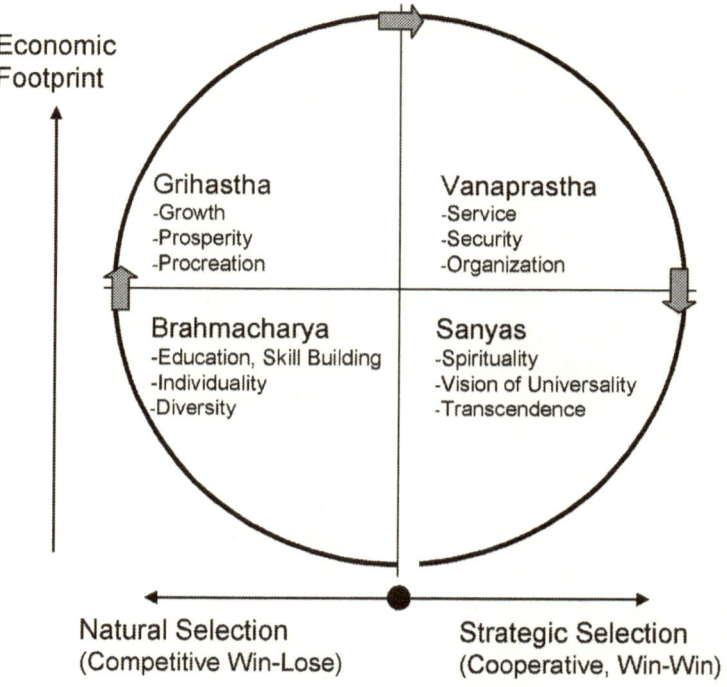

Figure 1.1: Mahashrama Cycle for humans

Figure 1.1 depicts the Mahashrama Cycle as it applies to humans. We start out life at the bottom center, and move slowly clockwise, growing both in our competitive ability to survive in this world, as well as slowly developing our economic footprint. This stage is known as Brahmacharya. The word Brahmacharya is derived from

two Sanskrit words – Brahma (meaning the totality of what exists) and Acharya (proficiency, or mastery of). The expectations are indeed high in this segment, as the budding human is expected to dedicate themselves to learning about the reality around us, and mastering those skills that would allow them to successfully move through the future stages of life as depicted in the Mahashrama Cycle. In addition to understanding our external existence, the student in Brahmacharya is also expected to develop more of their unique personality (or individuality) – and inculcate a good sense for who they are as a person. The unique perspective and skill set that we develop during the early years is what marks us as the individuals that we are – and collectively gives rise to the diversity that we enjoy within the human civilization.

The next stage of Mahashrama is Grihastha. The word Grihastha originally meant 'householder' – or someone who takes upon themselves to create a new family unit and work for its economic benefit. In today's environment, we have deconstructed the elements of Grihastha to mean Economic Growth and Prosperity, as we go about raising our next generation (Procreation). The family environment of ancient India was quite different from what we have today, but the task of raising our next generation and providing for them through economically meaningful activities remains the same.

The third segment is Vanaprastha, and it typically starts when we are done with the major duties of raising our children and securing our economic future. At the very top of the Mahashrama Cycle the Economic Footprint is at its peak, but the motivation to keep growing our individual wealth diminishes - as we begin to look at the world at large and try to figure out what causes are really worthy of our continued effort. In the olden times, the accomplished householder would literally go to the forest (Vana) to reflect and meditate, and come back energized to apply themselves to one or more areas of service to humanity. Part of the service motivation, of course, is to make society and its institutions more secure, organized and functioning effectively.

The final segment is Sanyas – or the realm of Spirituality and Transcendence. This is the domain where we take our material existence and cast out everything that is peripheral to our core identities and values. Spirituality is the art of searching out the vanishingly immaterial essence of our existence (our soul) and the core purpose for our existence. Transcendence relates to our ability to take the core ideas from our life and being able to apply them towards the concept of a far greater level of existence. The Vision of Universality is our conception of how we can build on our rather limited existence and extend it out to include the much greater existence that we conceived through our Transcendence. The Universality part comes from the desire for our existence to have meaning in the context of this Universe – which encompasses all intelligent life within it. For many of us, religion happens to be the regimen that helps us visualize our Transcendence, and gives us the way to relate to the ultimate purpose in the Universe – namely God. However, the concept of Sanyas does not need a specific religion (or any religion for that matter), as we seek to relate ourselves to the essential purpose of the Universe. The name Sanyas, also from Sanskrit, means the dissolution of the Self – and becoming one with the essential purpose of the Universe (or God).

The rise and fall of the economic footprint is easy to visualize – as our ability (and motivation) to work goes up and comes down again as we age. The left half of the Mahashrama cycle depends on competitive skill building (Brahmacharya), as well as a competitive garnering of resources for ourselves and our families (Grihastha). Hence, to a large degree, the left half of the Mahashrama cycle is subject to the laws of evolution and the principle of 'survival of the fittest'. By the time we enter Vanaprastha, we are now willing to work for the common good and the evolutionary fitness of our personal biological DNA is no longer our highest priority. We are now able to work on strategically important issues that we care about, without concern as to whether the benefits accrue directly to us. This cooperative existence is in sharp contrast to the 'Winner Take All' approach at the other end. We will call this cooperative process 'Strategic Selection' – where we focus our efforts on strategic goals with the potential to benefit all of humanity.

Relevance of Mahashrama in Today's World

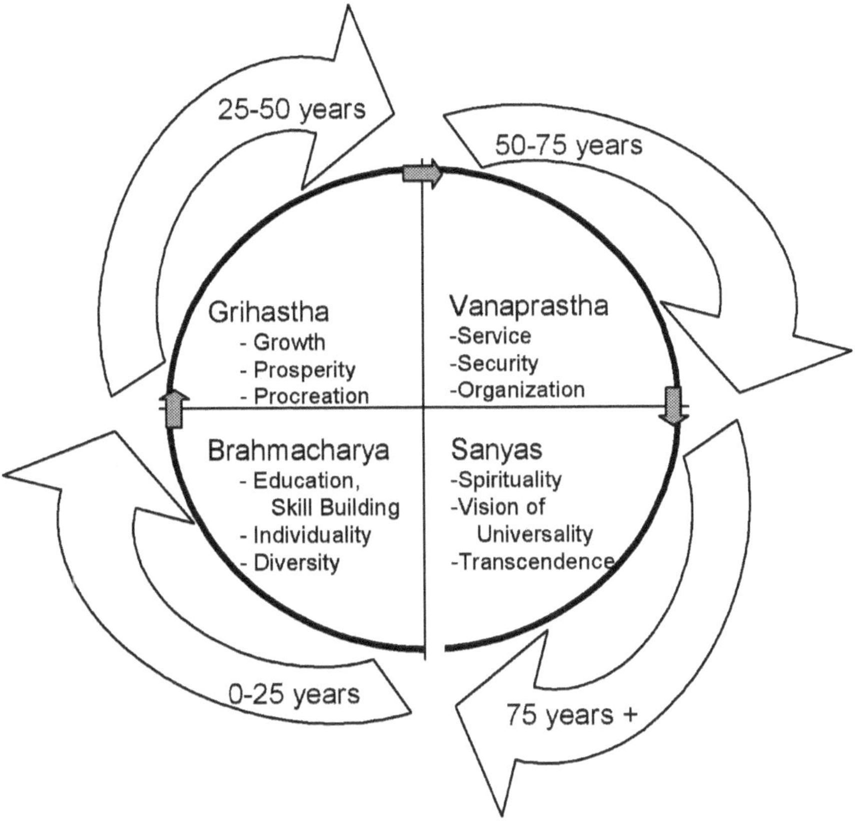

25-50 years

50-75 years

Grihastha
- Growth
- Prosperity
- Procreation

Vanaprastha
-Service
-Security
-Organization

Brahmacharya
- Education,
   Skill Building
- Individuality
- Diversity

Sanyas
-Spirituality
-Vision of
   Universality
-Transcendence

0-25 years

75 years +

Figure 1.2: Approximate Human Age-span along the Mahashrama Segments

If we go by age as a metric, it typically takes a person till the early to mid twenties to complete their education. Hence, conservatively, the Brahmacharya stage lasts from 0-25 years. In the mid to late 20's is when we feel established enough to start our own family, and typically takes us to about 50 years for us to raise our children, and help them become successful educated human beings in their own right. This is the age of the

Grihastha. Once the children are old enough to leave home, the individual may continue to work well into their 50's or 60's – but now will start having much more time for introspection and dedicating themselves to work for the common good. Hence 50-75 years is typically the age of the Vanaprastha. This is the age group of people whom we can usually count on to rise above the petty differences, and operate for the benefit of a large group of humans. These are the leaders who lead through service, and lay down that path for society to function in a secure and effective manner.

After the age of 75, there is typically little expectation for the individual to work in an economic sense, either for themselves or for their families. This is the age of Sanyas or spirituality. Contrary to popular perception, where we tend to categorize the spiritual domain as intensely private and not particularly beneficial for society at large, we will see in that Sanyas is a very critical segment for future civilization growth. The ability of those with the Sanyas mindset to operate in a plane well above our daily existence allows them to perceive long term strategies that can be a significant benefit for society. In ancient India, the segments were each designated to be 20 years – and it is entirely possible that future humans might stretch each segment well past the 25 years indicated in Figure 1.2. Let us now take a deeper dive as to what makes each segment unique and vital. As we go through each segment, please keep in mind that one segment is not necessarily better than the other. All four segments, as well as an effective set of elevators to move between the segments, will be needed for humanity to survive, progress, and prosper.

Brahmacharya

A person is born, and spends the very first part of his/her life learning how to become a successful human being. This phase of life we have referred to as Brahmacharya. The person undergoing this phase is called a brahmachari. As one goes through childhood and adolescence, one acquires a lot of knowledge and practices a lot of skills, not all of which will necessarily be needed in adulthood. At this stage, the individual is a net consumer of resources. Also at this stage, one hones their competitive skills, which often manifests itself via sports, games and other competitive activities. During this stage, the capacity of the human mind to learn is indeed immense – and none of the later phases come even close to the sheer volume of learning that happens during Brahmacharya.

In Indian philosophy, the competitive spirit was restrained and directed under the influence of a 'Guru' or teacher – and the student was expected to live a 'chaste' lifestyle. In adapting this concept for the worldwide human society – we will put a lot of emphasis on the Competitive skills – which, as noted before, are a critical component of the Market Economy. We will also move the focus away from the lifestyle choice of "simple living, high thinking," that was prescribed within the Ashrama system. Mostly, a brahmachari does not get to choose the circumstances under which they are born and raised. It is more important that they concentrate on building the capabilities and character that will make them successful long term – rather than fighting to change the circumstances that they find themselves in the short term.

The move away from requiring one to live a 'simple lifestyle' proposed for the original Brahmacharya, and the inculcation of market driven economic forces does not change the guiding purpose of the brahmachari. Their primary task still remains - to master the knowledge and skills that would set them up for success in the subsequent stages of life. Hence the Mahashrama cycle does not deviate from the intent of the original Ashrama cycle – but just brings it up to date with the reality we see today.

The competitive reality of the market economy is not the only reason for building in a strong base of competitiveness within Brahmacharya. Competitiveness is manifested at the biological level as well. It is a little known fact that the newborn human baby has many more brain cells (neurons) than the adult human being. Even the synaptic connections within the brain appear to max out at around age 2-3 years[1]. It appears that we learn more by a process of elimination than by building new neurons in our brain. A high degree of competitiveness is built into the process of evolution as well:

"There are millions of different species of animals and plants on earth -- possibly as many as 40 million. But somewhere between 5 and 50 billion species have existed at one time or another,"

- David Raup in his book "Extinction: Bad Genes or Bad Luck?."

This translates to less than one in a thousand species that ever existed is alive today! The battle for survival in a highly competitive environment is indeed intense, and it is important for the brahmachari to shore up on their repertoire of competitive skills.

All is fair in Love and War …

Like most creatures, human beings are at their most competitive behavior when it comes to the selection of mates. It is said that 'All is fair in Love and War' – suggesting that humans will do almost anything to gain an advantage in either field of endeavor. This competitiveness particularly affects the males of our species as is borne out by the following statistics. These numbers are based on US demographics, but similar trends exist worldwide.

- For every 100 females conceived, there are 115 males conceived
- For every 100 females born, there are 104 males born
- By age 25, women are in the majority, and enjoy significantly higher life expectancy.

The difference at every stage is higher mortality rates among the males, with a peaking effect in adolescence and early adulthood. The competitiveness of males takes many forms – including, in many cases, violence against each other. The demographics of violent offenders and other reckless actions also support this general observation.

Societies throughout the ages have exploited these natural tendencies. Men, in general, have been more expendable than women. Through most of history, men were allowed to have more than one wife – not because they were intrinsically better – but

because throughout history (and presumably prehistory), with its many ongoing feuds, there were much fewer men to go around, and population levels had to be maintained!

… and in Business

By extrapolation, the strength of ultra competitive markets, and the human competitiveness that drives it – will naturally yield high flying solutions that have little margin for error. We will see later that the long term survival rate of any particular business is comparable to the species survival rate – only enacted in a much faster timescale. By extrapolation, the products that corporations put out might be very attractive – but are intrinsically not the safest alternatives. As has happened with seatbelts and airbags – society (e.g. the governments in countries like the US) had to get together and mandate that these safety factors be made available on a certain timeline. This influence comes more from a Grihastha and Vanaprastha viewpoint as we will see later. Competitive behavior is not by itself sufficient to bring about safety and security - hence the need to rely on the later stages of the Mahashrama cycle for guidance. A successful Brahmacharya stage, in addition to giving us a degree of self sufficiency, also prepares us for the subsequent stages. Just like the first stage of a multi stage rocket ship – the character and skill set that we build here must carry the remaining stages of human endeavor upon its shoulders.

Grihastha

At the end of the Brahmacharya phase, and presumably with the selection of a mate – begins the economically most productive period of life – Grihastha. The individual who has made the successful transition to the Grihastha stage we will refer to as the grihasthi.

Grihastha, as a Mahashrama segment remains little changed from the time of the ancients – and is just as applicable today as it was then. The young adults have a lot of energy, which they devote to economic activity, as well as to the challenging tasks of raising a family. The initial competitiveness slowly begins to fall off – as the grihasthi now begins to think more about the family unit

than their own self. Our economic activity – which we will refer to as the economic footprint – continues to grow until we are well into our middle age, and our children, in turn, become capable of supporting themselves. This is the age of prosperity, which does by its own nature reduces the hunger for competitiveness, and over a period of time causes the economic footprint to max out.

The interdependence of competitiveness and economic footprint can be visualized as a half circle as it goes through the initial increase and subsequent decrease in competitive 'Win-Lose' behavior, while at the same time increasing the economic footprint. Brahmacharya and Grihastha together constitute the left half of the Mahashrama Cycle shown in Figure 1.1. To a degree, these first two stages can also be applied to any animals that tend to their young, often at significant self sacrifice.

For animals that do not expend any effort into raising their children – essentially finishing their parental duties when the fertilized eggs are laid – the Mahashrama Cycle operates only in one quadrant. Most fishes, amphibians and reptiles fall in this category. The young in these species are capable of fending for themselves reasonably well upon birth, and are usually numerous enough that a large fall-out throughout their infancy and juvenile stages is part of the equation. Even amongst birds and mammals, it is often just the mother that takes care of the juveniles until they attain a certain level of independent existence. Sadly to say, even amongst humans we can find niches in society where the male contribution towards raising our next generation in Grihastha is tragically missing.

In all of society, there is one heady transition – the one between Brahmacharya and Grihastha - which we seem to agree on and even idolize and worship. Youth and vigor, as well as the competitive ability to 'win' appears to hold a favored place in our collective psyche. Closely related to the concept of winning, and taking the first step towards Grihastha – is winning at romance. Too often, this emphasis on competing and winning takes the form of sex and violence – which all too frequently permeates cultural expression ranging from cinema to music to even video games.

Thus, in many ways, the successful completion of the Grihastha stage and the potential for passage into Vanaprastha is what really differentiates us humans from our animal counterparts. This is not to say that there are no acts of kindness or altruism found in animals, only that such acts can usually be placed in the context of survival benefits for the individual or their extended family or tribal unit. Symbiotic cross species relationships do exist in nature – but these can also be demonstrated to be useful to both parties involved. It is indeed ironic that in evolving to a bipedal animal – humans gave up the use of two limbs that could be used for locomotion. What their freed up arms and their brain has earned them is the capacity to operate in all four segments of the Mahashrama cycle – while our animal cousins (including the great apes) continue to labor under the propulsive power of only two Mahashrama stages…

Vanaprastha

When the responsibilities of Grihastha are completed, and the children are well on the way to becoming grihasthis themselves, there comes about a very profound change in our outlook to life. We are just off of the height of our economic footprint, and our personal productivity and motivation to work for a living are showing indications of winding down. We have been through our mid-life crises, and come to terms with the fact that the most productive years (biologically and economically) are behind us. This is the time for deep introspection, and figuring out the causes that really bring meaning to our lives. Once we have identified a greater purpose, this is the segment where we apply ourselves through service leadership to bring about the changes that we seek around us.

It is important to note that even if we make the changes economically, mentally a lot of people have the hardest time making this adjustment to Vanaprastha. Especially in the economically highly competitive societies (e.g. in the Western countries), often we will see people clinging to behaviors from the past phases of life. This emphasis on youth and vigor and competitiveness has economic benefits – but it also has undesirable side effects. Without a timely transition into

Vanaprastha, there is not enough social energy spent in the activities that lead to a greater level of security and organization for society as a whole. The Vanaprastha leadership is not one of hierarchical order giving, but one of service and applying oneself to the areas where the need is the greatest. We will refer to the brave souls who have successfully completed the transition to the Vanaprastha stage as the vanaprasthi.

In India, during ancient times, when people were much more in tune with nature, Vanaprastha literally meant – resident in the forest. The idea here was that the ex-Grihastha would spend a lot of time in the forest in meditation and penance, and inculcate a vision of the greater scheme of things in life. Through this self reflection, they would come back motivated to work on the things that really mattered to society at large.

For the first time in our life, we are comparatively free of the prime biological source of motivation – i.e. the need for our individual genome to procreate and create viable offspring. Now, as we look wider, we realize that we are part of a much bigger cycle in life, and that there is a lot of common human interests worth working on. For some people, it might be casting the nets just a little wider and caring for their aging parents, relatives, or friends. Other people will take it further still and want to be of service to their own community or religious group. Still others will go further out and want to do what's best for all humanity – like volunteering to teach the underprivileged in a far-away land. Going even further, some of the Vanaprastha, seeking to be of service to all of Mother Nature, might decide to dedicate themselves to causes like preserving the diversity of plant and animal life on planet earth.

In modern society today, the time and energy spent here on an individual level is relatively small (e.g. charitable activities), but is huge at the societal level. All of Government falls within this category. Similarly a host of social and cultural institutions that do not espouse the profit motive would fall under the Vanaprastha or Social Service umbrella. Although the institutions are based on social service, the individuals that make them work are often career staffers who will have very different internal motivations. Vast amounts of resources are spent in this area, yet we are often

left with the feeling that the people in positions to be of 'Service'
can hardly relate to the problems of the people they serve. This is
but one example of an individual-segment dissonance that could
limit the productivity of any segment of the Mahashrama Cycle.
This dissonance is especially harmful in the Vanaprastha role, as
these are the people entrusted with the well being of society as a
whole.

In Ancient India the 'Enlightened' Vanaprastha would often
become the 'Guru' and take upon themselves to teach and
guide the impressionable brahmacharis. It is easy to see why
the emphasis would be on austerities and 'Simple living, high
thinking'. In today's environment we often have the situation where
the young professor or school teacher is hardly much older than
the student. This may be fine from the standpoint of propagating
knowledge – but does little to illuminate the 'context' or overall
utility of the information (or content) being passed on.

To complicate things further, nowadays technology is moving so
fast that a huge 'Generation Gap' often exists between the parents
who are inching reluctantly towards their Vanaprastha stage, and
their children. The new generation may not even acknowledge
that their parent's experiences and viewpoint would be relevant to
their own specific circumstances. Even worse, the Grandparents
who might be well into their Vanaprastha stage, with plenty of time
on their hands, are unable to help out because their viewpoint is
considered totally irrelevant – not only by the grandchildren but
also by their parents!

The situation today is that most people would like to learn
from those that are just a bit ahead of themselves in the road
of life – hence setting the stage for huge exaggerations of the
characteristics of the phase they are in. It is little wonder, then,
that people have the hardest time leaving behind the first two
stages of life – the competitively enhanced Brahmacharya and the
productive/consumptive Grihastha.

Fortunately for us, there are several trends that herald the
resurgence of Vanaprastha principles into today's society. Yoga
– or spiritual joining – which is the ancient practice of empathic

co-existence is gaining popularity. Demographic trends, with the projected extension of life expectancy worldwide – suggests that there will continue to be a population boom in the Vanaprastha age group. Also, more and more people are retiring with considerable life's savings, and are not reluctant to use part of it to further causes that they feel are for the 'Greater Good'. The time is ripe for a discourse on 'what in life has value beyond competitiveness and asset accumulation?' The purposeful thinkers in this space will largely determine the long term success and viability of our human civilization.

Sanyas

The transition from Vanaprastha to Sanyas is one of disengagement from our material existence. In ancient times, the aging vanaprasthi came to the realization that the time for Service Leadership was getting over, and that one needed to come to terms with one's impending demise. The vanaprasthi who has been deeply immersed as the social worker – coaxing out the best individual and collective effort from society – can no longer continue to be the 'glue' that holds society together. The time has come for even deeper self-reflection and a renewed focus on those aspects of our existence that really brings meaning to our human experience. The emphasis on working for the common good slowly begins to wear off – heralding the beginning of the Sanyas stage. We will refer to the person who has completed a successful transition to Sanyas as the sanyasi.

Sanyas is the realm of strong religious and spiritual beliefs, and coming to terms with the 'utter simplification' – what of our diminishing senses and capabilities is the essence of what we really are as a person. For the previous three stages, the primary reward mechanism has been the fruits of our own labor – but with Sanyas we really have discovered the bottomless well of spiritual happiness which we will refer to as Ananda. We might still continue to work, but the gratification that work brings is no longer the source of our feeling of self worth and self fulfillment.

In these days of diminishing Social and Cultural diversity (i.e.
true diversity of viewpoint) it is becoming even more crucial
that we preserve the 'sublime essence' of what a 'way of life'
or thinking is all about. The word 'Spirituality' is often used in
this context – signifying a special connection to the 'Spirit' or
vanishingly immaterial essence of what it means to be ourselves.
The practical side of Sanyas, then, is to make sure we preserve
as much of our unique viewpoints and cultural perspective as
possible – in a way that is beneficial to future generations.

It is important to distinguish Spirituality from Religion. It is possible
to be utterly sincere in our Spirituality even if we do not believe
in a specific God or path to salvation as prescribed by a specific
religion. Spirituality refers to the 'Spirit' or the essence of who we
are, and the purposes we strive for. Religion is a set of beliefs,
often held in intricate detail by its practitioners, as to the nature of
'God' and the path to 'Salvation' or union with one's own God. The
fact that spirituality and religion seem to go hand in hand should
not detract from the potential for each to be consistent in itself
without need for the other.

In a more practical vein, Sanyas is the realm in which we can
cast our nets wide and think through the various scenarios of
human primacy, and our ability to bring God-like powers to a
future version of human existence. Even if there were no God
that fits the Omnipresent, Omnipotent and Benevolent entity that
many of us imagine – could we go about creating such a supreme
presence in our little segment of the Universe? If we were thus
capable, could such a super intelligence find a way to expand to
fill the universe as we know it? If there were a being or intelligence
that had already accomplished such a feat, what would be their
expectations of us as an emerging civilization? We will see later
(Chapter 6) that an integrative framework does exist that would
allow humanity to march together to greater glory - whichever
way the great unknowns pondered here might end up getting
expressed.

Yes, there is a distinct possibility by which no Civilization or
Intelligence before us has proceeded to develop the qualities
that we associate with God – and we humans will be fortunate

enough to develop such qualities on our very own! To unlock this potential, we will need to operate well in all four Mahashrama segments, and thereby find a way to take the 'gems' of wisdom from our current existence and use these to transcend ourselves into an immensely higher plane of existence. This harvesting of 'gems' or 'Accumulating Cultural DNA' for future generations is a key role of Sanyas. Another key role arising out of Sanyas is the 'Vision of Universality,' which seeks to connect our rather delicate existence with the intense possibility - that what we are about to become does really make a difference in this Universe! Unlike any other time in history, the tools and the technology are now at hand for us to lay out this glorious vision, and back it up with some detailed planning. We will find in this book that many of the technological and philosophical underpinning for sustained human civilization growth is currently actionable – and that our individual and collective decisions over the next hundred years will largely determine how high we can fly as a civilization!

Magnification of the Mahashrama Cycle

In progressive times, the Mahashrama Cycle increases in magnitude. The brahmachari learns more, and is exposed to more ideas from which he or she can select the most productive ones (which is a measure of competitiveness). In the transition to Grihastha, the brahmachari carries through more skills and proficiencies to help them build a good sized economic footprint. The resulting prosperity allows them to bring forth even more viable offspring – thus resulting in even larger cycles of human potential. In time, the successful grihasthi move on to Vanaprastha and devote their energies in organizing and securing our collective efforts as a society. The sanyasi conceives an even bigger vision of success for our expanding civilization, which is now used by the vanaprasthi to set up the social institutions (e.g. education system, social services) that would slowly but surely lead us to the attainment of success with the stated Vision. This scenario is illustrated in Figure 1.3, and can lead to a series of expanding Mahashrama cycles.

We have referred to the key learnings that are crystallized out of the Sanyas stage and fed into the 'reality' of the ensuing Brahmacharya stage as our Cultural DNA. Just as surely as our genetic DNA defines who we are as individuals, our accumulating cultural DNA will define our identity as a civilization and our overall potential for success in this universe. When the expanding Mahashrama Cycle leads to the attainment of potentially God-like capability for humanity – the Cultural DNA will take the place of our genetic DNA to define who we are as the human civilization. Abstracting out the genetic DNA piece will make it possible for both our biological and machine progeny to identify with the human Cultural DNA as the basis of who we are. It may sound strange referring to our machine creations as 'children' or 'progeny'. Yet, once the cultural DNA becomes the determining factor of who we are – the machines with their superior capability to be coded with this cultural DNA will hold just as much right to be called our Children as our biological offspring. Hence, 'Nurture' becomes the 'Nature' for our next level of existence – and intelligent machines will have a key part to play in our civilizational success story. The key design requirements for developing our machine creations into fitting counterparts for our biological progeny are outlined in Chapter 9.

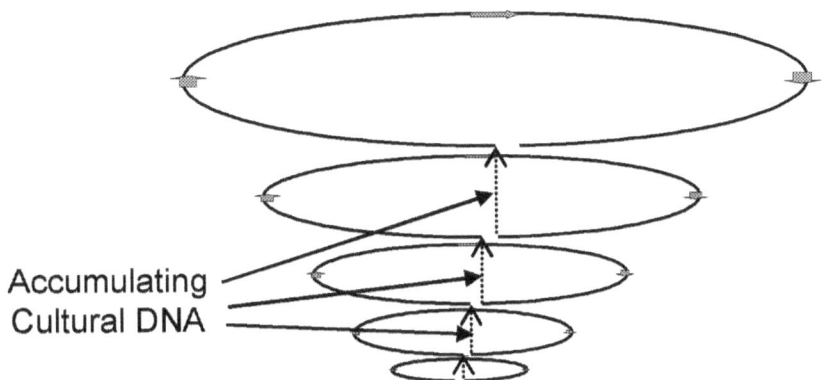

Figure 1.3: Expanding Mahashrama cycles with Accumulation of Cultural DNA

Typical pitfalls

A typical trap that humans tend to fall into in the face of prosperity is that they begin to neglect the competitive skill building required to expand the left side of the Mahashrama Cycle. Too often, we find out that economic success is used to bolster our ability to consume, and does little to equip our children with more forward looking competitive skills to help them maximize their potential, and through them the potential for our future human existence. Similarly, quite often we fail to internalize that our legacy as humans is critically dependent on a functional Vanaprastha, and don't end up investing enough in the glue that organizes and secures our society. Lastly, we often find ourselves shying away from deep discussions as to our essential identities and purpose for existence – as if we were powerless to make a determination of our own future success. In doing so, we find ourselves missing the integrative framework that can unify our efforts as we conceive of a collective glory that can only be achieved through concerted activity. When we observe such symptoms in the Mahashrama context, what we perceive is that the four segments are out of balance – and that not enough attention is being given to the natural flow of individuals, ideas and resources between the segments. Much of Chapter 4 is dedicated to identifying and satisfying the core human needs that are fundamental to the successful operation of the Mahashrama Cycle.

Applicability to Human Institutions

One of the key themes of this publication is that the same Mahashrama Cycle has applicability towards a range of human institutions. Multiplied a few thousand-fold in scale, it will apply to the Corporation – as we will see in the Chapter 2. Multiplied a few million-fold, it will apply to governments and nations. Multiplied a few billion-fold, the whole human civilization could benefit from a proper understanding of the Mahashrama Cycle.

Down the road, when we get to an intelligence capability of about a 100 trillion fold that of a single human – it can be projected that a new level of intelligence emerges, which we will refer to as the Supra Human Entity or 'Supra' for short.

Grihastha Vanaprastha
Brahmacharya Sanyas

As noted earlier, such tremendous levels of magnification would not be possible by organizing biological humans alone – and machine intelligence has to be brought on line in a systematic way to make this vision a reality. This is the same level of magnification as between the individual human body cell – and the collective of about a 100 trillion cells that makes up our human body. When we get to determining the inner workings of the Supra, we will see that many of the principles that govern the efficient working of the human organism also come to play at that tremendous level of amplification. Hence, the title of this book – In Our Own Image. In your author's estimation, this is the minimum level of capability that we would call 'God-like' – and could be measured by the ability to put together an entire planet, biosphere and fossil records and all – one single atom at a time! The magnification of the Mahashrama cycle does not end with the Supra. The scale of the universe is such that we could multiply the Supra level of intelligence another quadrillion-fold - and still not match the full scope of our Universe.

There is indeed a lot of room for growth! It is important that we start with the right practical mindset - so that our collective human 'essence' will continue to live and thrive within the DNA of the much greater Intelligence(s) to come. The search for 'Divinity' is inherent in our quest to transcend our rather meager existence into one that has some of the attributes that we associate with God. This is the positive side of the potential that we find within us as humans. If we do not heed this potential to transcend our existence, history has a very great likelihood of repeating through the cycles of destruction and renewal. Not only can that prevent us, as humans, from reaching our true potential – but we may inherit a fate in some way worse than death; i.e. of becoming inconsequential as a race and as a civilization!

Chapter 2
Mahashrama Cycle – Applied to Corporations

In life and business, there are two cardinal sins... The first is to act precipitously without thought and the second is to not act at all.

- Carl Icahn, Investor, Entrepreneur

Before we tackle how the Mahashrama Cycle applies to the rise and fall of civilizations, let us take a look at how it applies to a much faster moving human institution – the corporation. The corporation is the most powerful economic engine that we have in modern society. It is the products and services put out by corporations that we rely on for a large part of our day to day needs. The ability of the Corporation to have a global presence, and yet be able to be run by a few senior managers – gives it the ability to compete, react and even change itself in relatively small amounts of time compared to other human institutions. This quick 'beat-rate' and the ability to cope with rapidly changing market realities allows us to use the corporation as a 'microcosm' for Mahashrama learnings - especially as it applies to other human institutions that evolve at a much slower rate, like nations or even entire civilizations.

We are used to the idea that most corporations operate for-profit, and that there are also a relatively smaller number that operate on a non-profit basis. Interestingly enough, there is a third kind of corporation that have burst onto the scene with the age of the internet. These companies operate in the white space between the two traditional corporate setups. Here the primary services are offered for 'free' to anybody with access to the internet (e.g. electronic mail, search engines) – with the corporation earning its keep by delivering supplemental information (e.g. advertisements) often disguised as part of the packaging. This becomes possible because internet technology has now enabled the delivery of services to the end user so efficiently that for-profit companies can behave like non-profits, and give their services away for free! Let us call these companies supplemental-profit, which together

with the non-profits will have a significant role to play in the creation of the Vanaprastha service network of the future. The two advantages that these internet based companies have is that their beat-rate of innovations is super fast - and their ability to service just about any human on this planet verges on the omnipresent.

The beauty of the corporate setup, whether it is for-profit, non-profit or supplemental-profit - is that it allows a few relatively ordinary individuals to lead an organization that can have global impact. Due to this remarkable amplification of the qualities of the individuals in leadership positions within the corporation – a tremendous amount of effort goes into equipping these folks with the right skill set for corporate success. In this Chapter we will look at how leadership is interpreted through the lens of the Mahashrama Cycle, and how the direction setting process links our day to day activities to the larger reality and vision that drives us. Also with the great reach and efficiencies provided by our internet based electronic medium, a single enterprising human can now take on the trappings usually associated with corporations – like brand image, service reputation and strategic alliance building. Hence a lot of the corporate learnings can be applied just as easily to individual human beings.

Corporations and the Economy

As we search into ancient pre-history, the core concept of trading or bartering probably predates even the rise of a common language between the parties involved. Language is not even a problem today – all it takes is the ability to point and count on our fingers; as the author's family found out during a trip to the 'authentic' shopping areas of Shanghai and Beijing!

The basis of all economic activity is the transaction, or exchange. The invention of money just simplified the exchange process by providing standardized, easily portable unit of value. If you were a farmer, you could now concentrate on becoming a very good corn farmer or vegetable farmer, and thereby able to get the rest of your needs met by exchanging the corn or vegetables at the

'market' for other goods. But, does having the ability to exchange intrinsically create value? Or, does it just re-distribute existing value amongst the various players?

Having a relatively frictionless method of exchange does create value. Let us count the ways. First, we have specialization and economies of scale in production. The person that specializes, concentrates much of his/her efforts in one field of endeavor – thereby getting more output for the time invested than if he or she were Robinson Crusoe and directly responsible for all aspects of day to day survival. Beyond a single person, a firm that employs several people in a specialized area, gets even more economies of scale. Markets make economies of scale possible by providing a ready method of exchange for the goods or services produced.

Second, we have standardization of products and quantities by which they are traded. This allows economies of scale even beyond what the producers enjoy – a trader may buy grain from a hundred different producers, and sell it to a thousand different consumers – but he or she can do that only if there is some common quality attributes that identify these grains. The products exchanged are now called 'commodities' – which allow potential buyers to buy the product sight unseen. A day to day example is filling the car with gasoline – we trust so absolutely in the process of product delivery that we don't even see the product!

Third, it provides a matching of buyers and sellers, sometimes in locations spanning thousands of miles, and sometimes across time as well (e.g. the whole field of futures and options trading). With these types of economies of scale and global footprints, it is clear why mere humans, working individually - cannot tend to all the intricacies of businesses of such magnitude and scope. After all, we all need to sleep – even if for a few hours. The same is not true of corporations. It was once said that 'the Sun never sets on the British Empire.' Today, many global corporations can boast a presence in more countries around the world than even the British Empire did in its heyday. And tomorrow a small team of humans (or even a single human) with the aid of technology might have vibrant and responsive 24x7 operations in any part of the civilized world!

Mahashrama Cycle for Corporations

The key imperative for the typical corporation is the generation of wealth (profit) – which they do by providing the customer with goods and services that the customers value. The ones that are best at servicing the needs grow and prosper, and can accumulate great wealth and prestige. The ones that cannot compete effectively, see their customer base eaten away by 'fitter' competitors, and in the absence of protection, will cease to exist over a period of time. As we have noted earlier – the Market Based Economy that the Corporations compete within is the last major frontier in which the Darwinian principles of 'Natural Selection' still applies to human beings. This makes the Brahmacharya phase of the corporation especially instructional as a study of how we should set up human institutions for long term success.

Brahmacharya for Corporations

Figure 2.1 shows how the Mahashrama Cycle applies to a Corporation. The Startup phase of the business in a vibrant market economy will see companies being formed to utilize new ides, new technologies, or just re-targeting old ideas and technologies to a new customer base. Competition is usually fierce, and the mortality rate in this phase is quite high. The people that fund this critical activity are called Venture Capitalists, often unflatteringly referred to as Vulture Capitalists because of the very high expected returns and the high mortality rates involved.

> *66 percent of new establishments were still operating 2 years after they started in the second quarter of 1998, and 44 percent were still in existence 4 years after their birth. Findings do not differ significantly across industrial sectors.*
>
> *- Source: U.S. Small Business Administration, June 2004*

To illustrate this point, statistically more than 1 out of 2 new businesses started in the U.S. will be out of business in less than 4 years. This is a risky business indeed, and the way the

competitive market compensates for the high risk is by having even higher expectations of positive returns of the companies that do succeed. One of the things that the Venture Capitalists do is to clearly spell out who gets what in case things do not work out – which is a good thing to note even if the institution involved is something other than a corporation.

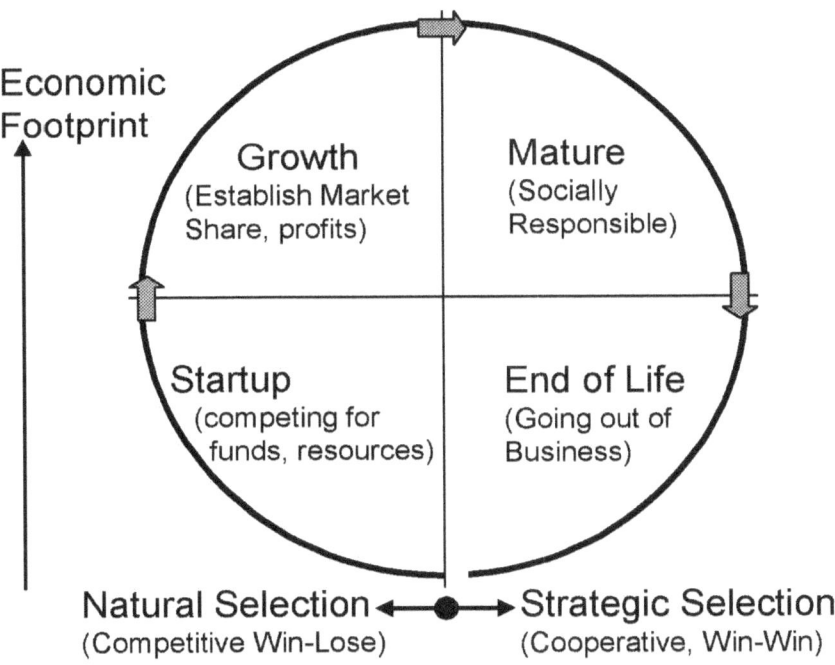

Figure 2.1: Mahashrama Cycle for Corporations

Grihastha for Corporations

The next Phase is the Growth Phase, where the products or services have been launched, and found sufficiently useful by the end consumer. The emphasis now shifts from survival to growth – and carving out the most significant customer base for the

product. Corporate Managers really like this phase, as a lot of fame and fortune can be earned with less and less risk of failing and going out of business.

Curiously enough, the person or persons that start the company and navigate it through the highly competitive survival process, are often not the best suited to lead the company through the Growth phase. Companies will often see one or more CEO (Chief Executive Officer) changes during the transition from Startup to Growth. Some of the CEO's who have just handed over the reins to the Growth phase CEO – will promptly go back and start their next startup venture. These people are often classified as 'Serial Entrepreneurs'.

Venture Capitalists will often insist that 'Professional Management' be brought in as the company gets ready to enter the Growth stage. If the CEO in the Startup stage is adept at juggling multiple balls, and motivating a team to put in tremendous creative efforts just to survive, the CEO at the Growth Stage has significantly different priorities. The first priority is usually to fill out the management team with people with a strong functional knowledge of Finance, Marketing, Sales, Manufacturing and so on. Once the team is in place, they will carefully chart an investment and growth plan that is robust and executable. Risk tolerance goes down significantly, and it becomes critical to be able to execute to a business plan that cements significant customer and revenue growth.

To borrow an analogy form the colonization of the American West – if the Startup phase was the Wild Frontier, the Growth Stage is the coming of the railroads and the shifting of a significant part of the population westwards. When a corporation is in the Growth stage - along come all the overhead of organized society - a Police force (Security), Communications/Computing Infrastructure (IT Department), a Legal System (Lawyers) and others charged with making sure that the growth plan can be executed with comparative safety and predictability. This is also the time to establish the Brand name of the corporation in the psyche of the customers, and a lot of effort goes into making sure that the Quality that the brand represents is well taken care of.

One of the reasons the corporations perform so well is that most of the people in responsible positions are themselves in the Grihastha stage. This makes it easy for their own personal life-outlook to be put to work - creating wealth and increasing the economic footprint of the corporation. Ironically, some corporate chieftains never mentally make the shift from Startup to Growth, and persist on doing things on the very edge of what's possible, or even what's legal. This similarity between how humans behave as individuals (extended Brahmacharya or competitiveness along the lines of 'all's fair in Love and War') and how corporations sometimes behave when they are led by such individuals – is an amplification of inappropriate human behavior at a massive Corporate scale.

It is to keep such tendencies from getting out of hand, that we need the thinking and institutions associated with the Vanaprastha Phase. The first line of defense is the company's own Board of Directors (BOD) and self regulatory arms like Accounting, Human Resources and Legal. If self regulations fail, it will be up to the society at large to come in and enforce laws of good corporate citizenship. If no prompt corrective action is taken, investors who can only tolerate so much risk - may actually decide that the corporate foundations is not strong or trustworthy enough to build their life's savings on, and quickly exit the stock market. It is just such an erosion of investor confidence that we experienced in the very early years of the twenty-first century.

Vanaprastha for Corporations

Let us take the following scenario to illustrate the 'mind-set' change that demarcates the transition from Growth to Social Service. Two car companies (fictitious, let's call them A and B) are faced with the same problem. Two years after their new compact sedans are introduced; both find a structural defect in the car that causes the car to blow its gas tank if stuck from the rear at a speed of 50 miles per hour or higher. To do a mid cycle design change (the same base model can run up to seven years) and to recall and fix all the effected vehicles will cost $500Million.

The company accountants pore through accident statistics, and figure out that this set of circumstances will actually happen approximately 50 times in the remaining life of the design. Even if there is a court case and the company is found liable, they can pay up to $10M per incident and still come out ahead. Current awards in similar lawsuits were averaging about $1Million, with another $250K in legal expenses. Economically, there would be little justification for the design change, so what is the company to do?

Company A in the Growth or Grihastha stage, when faced with this quandary, might decide that the risk is worth taking, and put in place a team of crack lawyers to minimize the company's financial exposure. Beyond the $500M actual cost, there is also the opportunity cost from the potential returns that the company would forfeit by not having invested the $500M in financially more productive activities.

Consider the case for Company B which has come to terms with the fact that growth prospects are rather limited, and deciding to service their existing customers to the best extent possible. The threshold of risk that the company would subject itself or its customers to is suddenly a lot lower. The company decides to pro-actively do the recall, fix the existing cars, do the design change and promote the safety record of its vehicles. This transition from 'Growth' priority to 'Social Responsibility' marks the entry into the Vanaprastha stage.

The above case may be a little extreme, but there are other subtler signs that the Vanaprastha mentality is setting in. The company becomes a lot more conscious of Quality, even more than what is required or expected. They start making contributions to local communities and national causes. Employees begin to see facilities like child-care and opportunities for volunteering made available to them through the company. All these benefits to the employees and the community come at a cost of course, which in the long term gets rewarded with greater employee and customer loyalty. The company almost seems to be saying – *we are here for the long term, so let us make our mutual co-existence as rewarding and fulfilling as possible*. The products will often lose their unique competitive advantages as the market segment

matures, but the relationships are now built - using which the company continues to harvest the business that it has established with its customers.

In the high tech businesses, this is also the time that companies will start paying out dividends – signaling that they now have enough funds to return some to the investor. A lot of the big names in technology, like Microsoft and IBM, belong in this category. This also indicates that they have the prime prerogative of growing their economic footprint under control, and are ready to become socially more responsible. The emphasis in this phase shifts to peaceful and beneficial coexistence and sharing of the prosperity.

Last but not least, there are those corporations whose primary purpose is to be of service. These corporations will typically incorporate as non-profit and their proceeds are typically tax exempt. The International Red Cross, Doctors without Borders, and even the Estate of Alfred Nobel who set up the Nobel Prize – would fall in the category of corporations that operate primarily in the Vanaprastha stage. As the purposeful Vanaprasthas of the 21st Century come together to conduct their business of moving humanity in a progressive, insightful manner – it can be expected that we will begin to see a lot more of these kinds of organizations in the near future. Coming to the rescue of the relatively slow moving non-profits, we have noted the ascendance of the supplemental-profit companies with their lightning quick reflexes and ability to reach individuals throughout our planet. These capabilities are expected to become the backbone of a Vanaprastha Network of the future – which would support and nourish the electronic presence of every single human amongst us.

Sanyas for Corporations

We have already noted that in the US, half of all new businesses started typically go out of business within 4 years. Even with the faster beat rate of market economics as compared to humans, this is a very frequent occurrence, and a lot of the lessons learned apply to individual humans as well.

A spiritual corporation must know from the very beginning what it is all about. These are sometimes referred to as the 'Values' of the corporation. A personal favorite of the author is Ananda. Ananda, or the spiritual joy of existence from ancient Indian philosophy, can be roughly equated to "Joie de Vivre" in French (Joy of Life). One example of valuing Ananda would be the inclusion of "Great Place to Work" as one of the key values. Another terminology along the same lines is "Customer Astonishment". It is the fostering of an internal culture that motivates employees to get up in the morning and look forward to accomplishing some meaningful work during the day – and the fostering of positive feelings with other stakeholders (customers, suppliers, shareholders) that come in contact with the corporation.

A natural outgrowth of the Values of the Corporation is the 'Mission' statement, which leads to Goals, Strategies and Tactics as we move progressively from the spiritual to the concrete. We can learn a lot about the values of a corporation by studying the 'Mission' of a company. Ones that emphasize the equity of the owners (e.g. Maximize Shareholder Value) will likely build their Value set to coincide with the prerogatives of the Grihastha or Growth stage. Ones that emphasize service to customers, will likely inculcate more of the Vanaprastha or Social Service values, and focus more on the benefits side of the balance sheet to the customers and other stakeholders.

The Legacy of Corporations

A great number of corporations are very active in the areas of research and development, and some of them (like the AT&T Bell labs and Xerox PARC) have been very instrumental in seeding the technologies that have become the underpinnings of modern existence. In many cases the parent companies are a shadow of their former self, but the overall contribution to the wealth of human knowledge is incredible, and remains forever imprinted in the psyche of our civilized existence. These are about ideas that can 'take flight' – and take on a new life sometimes far removed from its origins.

Even the business model that a company is built on could also take on a life of its very own. Take, for instance, the idea that a nationwide (or even global) network could be built on next-day package delivery. This idea was pioneered by Fred Smith of Federal Express – but there are quite a few companies today that have espoused that idea. Business model ideas are relatively easy to copy, but there is a whole set of ideas inside a company – often referred to as Intellectual Property (or IP for short) that also could have a life of their own. Some of these ideas are somewhat tied down as to who could use them (e.g. Patents, Copyrights, etc.) but there are also a vast number that go to the general pool of human knowledge, and will live on whether the company becomes a success (like FedEx) or becomes a grand experiment that failed to live up to expectations (like the Internet music distribution company – Napster).

Also, someone who starts a business once is more than likely to start a business again. The new business may or may not have the same goals as the one just concluded, but is likely to have a similar set of core values. Quite often, even apart from the core values, some of the implementation ideas will also be transported into this new body.

Liquidation and Bankruptcy

So what else is a company in the Start-up phase to do, knowing that chances are that it could be out of business within four years! Nobody likes to think about bankruptcy and liquidation, yet quite a bit of thought goes into it when a business raises financing from, say, a bank or a Venture Capitalist. Typically, the bank that has loaned money to the company gets first dibs on any monies resulting from liquidation, then the VC, and finally if there is any left - the common shareholders (or owners). Quite often, there is not enough to go around, and the owner gets nothing. Going into bankruptcy will usually release the corporation of any outstanding debts beyond what it can pay back – so at least the owner(s) do not lose anything more than what the put into the business.

The process may seem unduly macabre, leading to images of a bunch of vultures fighting over scraps – but it does not need to be that way if the eventualities are well thought out. There is a reason the banks want the highest level of security – because the money often comes from people who demand a high level of security from the bank itself. Before the government (in the US and elsewhere) stepped in and insured individuals savings from bank failures, many a life's savings was lost because a bank made a few bad loans.

Parallels can be drawn here between a corporation and a human being preparing for Sanyas. On the financial side, the intricate structure of banks and VC's liquidation prerogatives get replaced by the 'Will' and 'Trusts' and other mechanisms that an individual may set up. The primary legacy of the corporation, just like that of the individual – lives on in the minds and behaviors of the persons that have been 'touched'.

Mechanics of a Typical Corporation

The typical corporation, as the primary driver of economic activity, falls squarely in the Grihastha segment. Yet, it has a significant overlap into both the Brahmacharya and Vanaprastha segments. Figure 2.2 depicts the overall half circle that Corporations generally cover – as it maps on to the full Mahashrama cycle.

In any established corporation with established products, it becomes readily apparent that the centers of power residing with the CEO and the Board actually favor the high economic footprint and low risk portions of the corporate curve. Corporations that end up getting cornered in this space in the name of maximizing shareholder wealth will often lose sight of the far reaches of market reality and new product or concept development – and can be surprised by a disruptive technology.[1]

The main difference between a family run business and the typical corporation is the existence of a somewhat independent Board of Directors (BOD). The corporate semicircle does go well into the Vanaprastha stage, and that is the domain of the BOD. The BOD is supported by functions within the corporation that provide

some degree of transparency into the financials (accounting, finance) and appropriateness of corporate behavior (legal, Human Resources). Yet oversight is only really half of the responsibility of the BOD.

Figure 2.2: Flow of products and accountability in corporations

The other half of the responsibility for a BOD has to do with the parts of the Mahashrama cycle that are not covered within the typical corporate organizational framework. This is illustrated in Figure 2.3. From the early Brahmacharya stage come the effusive rush of new ideas and new ways of doing business – which represent the most promising new opportunity for the corporation going forward. Arising from a good grounding of the nature of the new competitive reality, comes the question – is the corporation in a good shape to take advantage of the changes? Or, does it need

to re-invent itself in face of the emerging competitive reality? A re-invention is nothing more than a realignment of the Values and the Core Identity of the corporation – which is an activity deeply embedded in the Sanyas stage. Once the emerging opportunities have been checked against the core values (and realigned if needed), we move on the Vanaprastha stage.

The new opportunities will typically bring benefits to consumers and society at large – and these potential contributions need to be articulated as the 'Value Proposition' for a proposed set or products or services. In most cases this Value Proposition directly translates to the economic potential of the products themselves – and is indicative of how well the corporation will do in the near future. The activity of identifying the benefits of the product or service is clearly rooted in the Vanaprastha stage. The Vanaprastha association is also clear for the related task of persuading other co-travelers necessary for the successful delivery of the product or service to come and join the party. Co-travelers, in this case, may mean suppliers, marketing channels (e.g. retailers) or even government institutions that might need to make a rule or two to enable the smooth functioning of an emerging market segment.

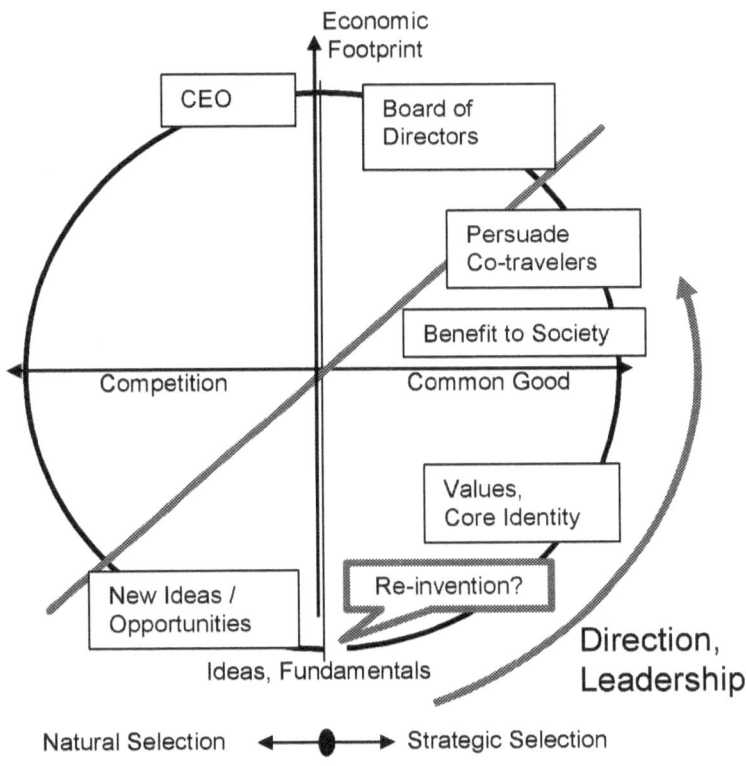

Figure 2.3: The BOD, and the other half of the Mahashrama Cycle

Together, this linking of the emerging realities from the Brahmacharya segment – all the way to enumerating the value proposition of the product/service and persuading other co-travelers to come along for the ride – can be summarized as the flow of 'Direction' that constitutes the inspirational leadership role of the BOD. Once the core ideas and their benefits have been identified, the normal corporate process starting with Market Research can take over – resulting in a fresh product cycle for the company. It is important to note that inspiration travels in the opposite direction to the regular clockwise movement of individuals (and corporations) over the Mahashrama cycle. In many cases, since the future can be known only as a juxtaposition

of several possible scenarios – the opportunities that are extracted from the Brahmacharya segment also need to be considered as an overlap of several scenarios.

The CEO and some of the very senior management may find themselves to be members of the BOD as well. However, there is good reason why the person who is the CEO, and hence responsible for the Accountability of the corporation, is not also the Chairman of the Board – with it accompanying oversight and "path finding" roles. But even as a member of the BOD, the CEO will have to strive to keep the balance between optimizing for the current environment, and sniffing out the greater context in order to refine and even redefine the corporation in its path to greater success.

Complexity Management

We have already seen how accountability flows up the economic curve for the corporation. For the Corporation, or any other hierarchically organized human institution (like the military) – two things need to happen. One, an incredible amount of action has to be brought to bear on the productive processes across the corporation – i.e. the corporation must take on the attributes of a well oiled machine. The second is that there needs to be one or more nerve centers, from which the intelligence comes as to which of the many tactical and strategic options available should be executed against. And of course, the well oiled execution engine needs to be connected to the nerve center so that the corporation functions almost as a living entity – constantly evaluating customer needs, competitive threats and market dynamics and reacting in a way that is consistent with its brand promise to the customer and the shareholder.

In the olden days, corporations used to be very hierarchical, and market and customer realities used to take a long time to make it to the people at the top that had the decision power. Information technology is changing all that. Today it is not unusual for a CFO to have the financial transactions to be captured and updated at the end of a week's transactions – or even a single day's activities! The trend is quite clear – as information technology works hard to

maintain 'just in time' inventory, and can be set up to provide an alert within minutes of any emerging trends or issues. As these systems get faster and faster, and the update times become closer to a literal split-second – the term 'nerve center' begins to take on an almost literal meaning.

The vision of a Mission Control center for a space mission comes to mind – with an array of displays and indicators all tracking hundreds of details about the mission in question. Decisions often have to be made quickly, and with incomplete data. Just like in the human body, the fastest response times are had - not by signals that are relayed to the brain, processed and responded to – but by semi autonomous reflex action. Decisions that require very short response times are typically delegated to a functionary who is very conversant with the details of what's happening – along with guidelines as to what the appropriate reactions may be for a wide range of possible inputs. Occasionally, a set of stimulus will come along that does not fit easily with the existing guidelines, and then these decisions need to be escalated to the next level.

For quick overall decision making and reaction to trends, corporations should have as few levels as possible of people aggregating exceptions from the level below them and trying to process a viable response. With too many levels and a lack of accountability for stimulus response – critical inputs can get utterly lost. One example that comes to mind is an FBI report (originating in Phoenix) generated several months before the September 11, 2001 attacks – noting a very high level of activity from possible terrorists in flight schools and warning that an attack may be likely.

If too many levels of aggregation are not good – too few levels may also be unworkable. Imagine if the CEO had to deal with all the details of every day to day activity – including stocking the office supplies drawer! Yes, intermediate levels of decision aggregators are needed – which roughly follows the management structure of modern corporations. One indicator of how well the process works is the 'span' of the individual levels of management. This is the number of direct reports that a manager has. With the proliferation of information technology and decision support systems, it is not unusual to have ten or more direct reports in a

mainstream organization. With this kind of span, there may be only two levels of decision aggregation between the CEO and the frontline worker for a corporation with a thousand employees.

There is also a trend towards managing the complexity of our own individual and family lives, which corporations are gearing up to address. We are already seeing an effort to provide 'One Stop Shopping' for medical services through health management organizations. We even have health insurance for pets at designated pet hospitals. We will often buy extended warranties for expensive purchases like cars and in-home maintenance contracts for major appliances. Over time, it is expected that service – whenever and wherever it is required – will become the norm rather than the exception. Yes, in tomorrow's complex world, complexity management may easily become a value added service in itself! The opportunities are endless as to what goods and services might be packaged together to provide the most compelling value proposition to the end customer.

The electronic worker – the hollowing out of the Work Force

If we look at where the human worker is most likely to be replaced, or at least augmented, by the electronic worker – a curious pattern emerges. The areas most at risk, at least initially, are the middle order of the work force. These would be predominantly white collar workers who generally take in information, process it, and put out a condensed and updated version of the information. To illustrate the point, most of our family's banking today is done at Automatic Teller Machines (ATMs), but when we want pizza, the delivery person is still very much a human worker. On the other hand, the jobs at the very top that require strategizing and deal making are still very much the domain of highly paid humans.

In many ways, the work of the pizza delivery person is more complex than that of the human bank teller behind the counter. We are still quite far away from Automatic Driving Machines – and the rules of the road are written with a human in mind. Similarly, when it comes to everyday human interface activities that we take for granted – like the receptionist at the front desk – these will likely continue to be human for a long time.

There is a great divide in the works as human workers end up in one of two camps – the interface to the real world (like the pizza delivery person and the front desk receptionist), and the strategist – deal maker. Several highly regarded professions also fall into the real world interface category – namely teachers and doctors. The real world interface in either case is other human beings. In the longer term, even these areas will come more and more under the purview of the electronic worker. But for now, professionals of what we call the 'Service Sector' continue to be somewhat shielded from the effects of the hollowing out.

The hollowing out of the human work force parallels the hollowing out of the Corporations that we will discuss later in this Chapter – but is a distinctly different phenomenon. The fact that both seem to be happening at the same time at the early part of the 21st century looks to be more of a coincidence than part of some grand design. Yet, the effects on our society and individual careers are likely to be very dramatic.

Long Range Career selection – the last frontier

When it comes to the careers of our children or grandchildren, what advise would we give them as to which end of the highly polarized workforce to shoot for? With the exploits of the electronic worker expanding to newer and newer areas of human endeavor, the beachhead that we really need to defend is the role of the overall strategic leader and deal maker. But, how many CEO's and CTO's does the world need – you might ask? The answer is – more than we might think.

The forces responsible for Hollowing-out of Corporations (i.e. the desire to focus on a few core competencies, and leave the rest to others) – now yield a host of new areas that then become the core competencies of others. The outlook for small outfits that are really good at what they do will get stronger and stronger. The fact that we can now use the electronic worker as a force multiplier of tremendous magnitude could actually allow just a few humans to control vast swathes of economic activity. The person or persons

who anticipate the needs of the future and prepare themselves well for the new competitive reality could become tremendously wealthy and influential.

The Makings of a good Strategist and Deal maker

So what are the attributes of a good strategist and deal maker? The number one attribute is the ability to simultaneously ground ourselves in the reality of the current environment, as well as get a good grasp of how the opportunities can be expected to change over time. We will call this Perspective. Next is the creative ability to connect the opportunities in such a way as to create a composite opportunity that is greater than the sum of the individual opportunities. We will call this the Vision. Third is the ability to get all the stakeholders together and convince them that the Vision is of benefit to all the players. We will call this Persuasion. And finally, the one that gets a disproportionate amount of attention today in management circles – is the ability to drive to action and results. This we will call Execution.

Vision, Persuasion and Execution are all staples encouraged in Senior Management worldwide – but it may all amount to little without the proper Perspective or Worldview. Today, there is a thriving community of analysts that cover different business areas – and corporations pay dearly to subscribe to their analysis of the opportunities worldwide. What is not evident is that there is a strong positive feedback loop built into the paid-perspective process. Analysts are typically seen to concentrate on certain existing industries, and are not particularly good at discovering the new opportunities that come out of the confluence of two or more trends. Too often they will be influenced by what they think the corporate chieftains bankrolling them would want to hear. Also, with the typical new CEO tenured for just a few years to prove themselves, the truly far-reaching opportunities are easily missed.

Leadership and the Mahashrama cycle

The term 'Leader' and 'Leadership' are very prevalent in corporate circles, but what does Leadership really mean? For starters, let us look at what leadership isn't. Leadership isn't doing and

thinking what everybody else is doing. It also is not being utterly conservative in life and staying within very rigid boundaries of thought and action. If we look at the qualities of the Strategist and the Deal Maker, a pattern begins to emerge as to what leadership is truly about.

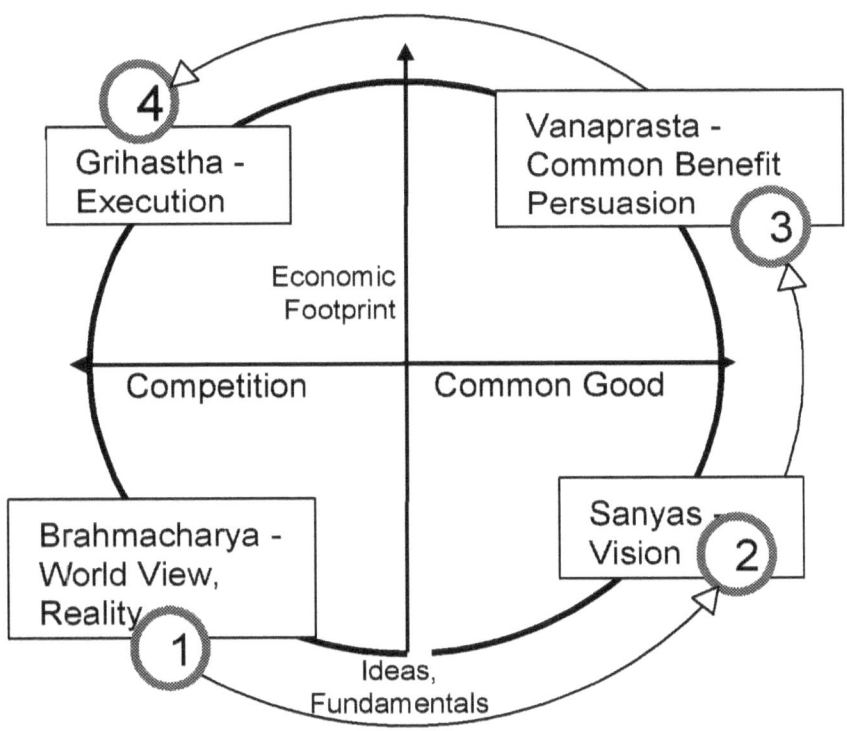

Figure 2.4: Strategic Leadership and the Mahashrama Cycle

One thing becomes immediately obvious in looking at Figure 2.4. The sequence of leadership flows in the opposite direction to the normal progression of humans through the stages of life! Once the leader has developed the appropriate level of Perspective or World View (from the Brahmacharya part of the cycle), they must take that and distill the essence of how a productive endeavor may be created. This is the Vision – and typically will include the elements of how a successful scenario plays out through time. The distillation of the essential/core purpose is what we have

come to associate with the Sanyas stage of Mahashrama. The next step is to take this Vision and convince all the stakeholders that such an effort is worth undertaking. This articulation of the common benefit or Persuasion is squarely a Vanaprastha activity. Finally we will take the common benefit that has been defined and turn it into Action and Results in the domain of the Grihastha.

Leadership v. Feudal Hierarchy

One might take a quick look at Figure 2.4 and perceive that this is a lot of effort to just get somebody to do something. Is it not much simpler if everybody knew their roles, and if a request came from a person at a higher level of authority – just to execute on it blindly? In the immortal words of Lord Tennyson, would it not be simpler just to follow the adage:

> *Theirs not to make reply,*
> *Theirs not to reason why,*
> *Theirs but to do and die.*
> > *- Charge of the Light Brigade, Alfred*
> > *Tennysion*

After all, entire empires have been built on hierarchical order-giving and expeditious execution! Even the whole idea of management and hierarchy in the corporation is based on the implicit or explicit assumption of 'obeying' a superior's command.

The answer is – today in the world of instantaneous communication and wide availability of information, one can manage a group of people using the tried and tested principles of hierarchy, but one cannot lead in such a fashion. It is possible that after working with a team over the full product delivery cycle, one could establish such a credible history of good Strategic Leadership that people on the team would trust us implicitly to have gone through steps 1, 2 and 3. Even then, good leaders will constantly reaffirm the full process of their thinking through the leadership stages to retain their good standing within the organization..

Perspective viewed through Time

As we have noted, the first step of Leadership starts with having a good grounding or Perspective on reality. Yet, any market or societal trend that we discern today takes time to go through the four steps and bear fruit in the form of a beneficial economic outcome. One of the characteristics of a Strategist is figuring out where the environment would be when the desired outcome can be delivered, as well has how things could change over the lifetime of the product or service. The target is definitely moving, and it takes great skill to shoot for where the opportunity will be when we can get there ourselves!

It has been said that technology projections regularly over promise in the short term, but undershoot what can be done in the long term. The most successful Strategic leaders will be the ones who understand the long term directions, and are able to formulate a short or medium term strategy that delivers a substantial portion of the benefits relatively quickly.

Virtualization (Hollowing out) of Corporations

With access to Global Markets, and the extreme economies of scale it provided, corporations quickly figured out that they did not have to build every part of the product themselves. In the early days of the Ford motor company, there wasn't an extensive supplier network, and it is understandable why the Ford 'factory' would take in raw iron at one end and put out Model T's at the other. Today a typical automobile may have an upgraded sound system from Bose, an antilock braking system from Bosch, and tires from Michelin. Hence, essentially the same car may be available under more than one brand name.

The term 'Vertical' is often used to describe the corporation that would take responsibility for every part of their production flow. Conversely, the term 'Horizontal' is applied to describe a corporation that specializes in only a narrow part of the total production or development flow – and makes that product or service available on a competitive basis to all customers. Yes, language, cultural and other barriers remain today in integrating

a product or service rendered from several continents. However, with the emergence of global quality standards and English as the dominant language for business transactions – these barriers are also disappearing.

The business case for going 'Horizontal' is very compelling, and lots of companies have undertaken extensive restructuring to find their 'roots' or core identity. Many companies have found out that they can make a handsome living just by specializing on those non-core activities that other companies were divesting. For example, Contract Manufacturers today will not only develop and manufacture your products, but will ship it to your customer – giving the appearance that it actually came from you! The Corporation that farms out essentially all of its development and manufacturing is often called a 'Virtual' corporation.

In its extreme, the Hollowed-out Corporation is left only with its name, and the value of the Brand – which is used to convey a sense of quality and satisfaction with an otherwise totally 'Virtual' product or service. It is very possible that in the future, Brand owners will have their Brands available for hire for those willing to create new products under the general parameters of the 'Brand promise'. To a limited extent this is already happening today on the Internet – with Companies like Yahoo and Amazon hosting multitudes of smaller merchants under their 'Shopping' umbrella – and providing some degree of legitimacy and quality expectation to the end customer.

The Brand

Now that we have established why the corporation is so critical to the functioning of the Market Economy, it is important to realize how it is that corporations project their value through vast distances. The number one intangible asset that global corporations have is their brand. The brand signifies a certain level of quality and satisfaction with the products and services; which corporations tend to reinforce with 'attractive' offerings and related advertising. Nowhere is this exemplified better than in the soft

drink industry – where the global producers like Coke and Pepsi provide minimally differentiated product to a worldwide audience based primarily on the value of their brand.

The brand itself may stand for more than the quality of a product or service, it can stand for innovation as well – especially in the high tech industry. Corporations pay a lot of attention to their brand, and are careful about cultivating a favorable perception amongst would be consumers. As the information industry matures, we have seen how all the other economies of scale will potentially become available to the small corporate outfit at reasonable prices. In the extreme, the hollowed out corporation will spend all their energy on maintaining the customer relationships and the perception of the brand – leaving others to do all the development, production, supply and logistics.

Quality Expectation from an Individual – our personal Brand

Venture Capitalists often have a challenge when they come across a new startup looking for an investment. The business plan and the general believability of the risk-return story is just the beginning. The CEO and senior management are put under a microscope and their past history of success and failure is pored over with a fine tooth comb. In many ways this is similar to the due diligence that a lender will do on us before agreeing to lend us money. The process is in place for a rather imperfect tool today in the form of the Credit Report for individuals seeking to borrow. There are research firms available for hire today that well do your research for you on the qualifications of a job candidate or a prospective CEO of a startup. This report, in a way, is your 'Brand promise'.

The Credit reporting process today, as a precursor of our individual Brand Promise, is deeply flawed as an instrument. Yet, it's essential utility lies in the fact that it tracks our public transactions record to paint a picture of our overall financial credibility. Similarly, a Personnel Dossier provided by a new-hire research firm paints a picture of one's potential to be successful as a future employee. The problem with both the Credit Report and the background checks conducted for employment is – one bad data point can spoil the utility of the whole exercise.

What's missing in the process today is the ability to get closure on the events reported. A future robust 'Public Report' will have the capability to capture the points and counterpoints on any contentious issues. It is also likely to cover a lot more ground than financial reporting – to the extent that very little of our public life would be truly private any more. If it begins to sound more and more like running for public office, where your opponent is looking to run a smear campaign – there are some lessons to be learned from such a parallel. From a very early age, one has to be careful to cultivate a Brand Promise that is appropriate for our future aspirations. This applies equally to personal ambitions as well as corporate entities seeking to establish themselves. In the extreme, with the hollowed out corporation – the individual and the corporation may become one and the same. We see many talented employees who leave their jobs and move on to found consulting companies of which they are the President and the only employee. We also see ambitious young entrepreneurs (sometimes even teenagers) making a business out of collecting and repackaging products that they sell through E-Bay. A certain amount of rating/feedback and 'Brand' building is built into almost any electronic marketplace that we see today – but the emphasis will only get stronger as more and more of the world economy starts to work electronically. Along with personal brand building, we will also need security and authentication services to make sure that we protect our hard earned Brand Identity from would-be scam artists.

Return on Investment – Objective or Subjective?

In the Capital markets, we are constantly calculating the 'expected' return and risks before, during and after we invest – as we do on an individual level. This takes on many forms – for example the projects we undertake, the jobs we do, the leisure activities we pursue, and even the relationships we build. Economists will often refer to this activity as maximizing the 'Utility Function'. Truth is, for most of us this utility function is so variable on past and recent experiences, or even our mood at a particular instance of time that it might as well not be a calculated 'Function' but a visceral and intuitive 'Gut response'. The principle still applies, however, as we seek to get the maximum return for

the investment (in time, money and energy) we put in. The return we get may be tangible or intangible. Depending on the quadrant of the Mahashrama Cycle we are in – the benefit may go to ourselves (Brahmacharya), our families (Grihastha), our society and fellow creatures (Vanaprastha) or our ability to simplify things and move on in life (Sanyas).

Magnification of the Mahashrama Cycle

Working through the parallels between how the Mahashrama cycle applies to our individual lives, and the highly amplified version of it that applies to corporations – we can conclude that they share the very same underlying principles. In fact, we have seen that with the added reach of technology, in many cases a single person can carry out the functions of a multi faceted corporation from just a few decades ago. The one aspect of the Mahashrama cycle that is expressed better in corporations (than in our individual lives) is the ability to re-invent itself in the face of a changing economic reality. The faster beat rate of corporations allows us to see the results of competitive positioning relatively quickly, and the need for re-invention arises on a regular basis. The companies that do well to change with their times, tend to retain the vigor to compete with the younger start-ups, and have a long successful track record. Ones that feel that they have found a success formula that they think would hold forever – usually tend to mature and then shrink to a shadow of their former selves. This message of a balanced application of the Mahashrama stages to corporations is equally applicable to other human institutions as well – including communities, ethnicities, nations and even civilizations. Our next focus will be the relevance of Mahashrama to governance – which relates to how large communities of humans organize together for mutual benefit. This organization can be at the local, national or even international level. At the national and international levels the decisions on how we organize ourselves for future success, will often have direct implications as to how successful we will be long-term as the human civilization.

Chapter 3
Mahashrama and Governance

All who have meditated on the art of governing mankind have been convinced that the fate of empires depends on the education of youth.
- Aristotle

I have gained this by philosophy: that I do without being commanded what others do only from fear of the law.
- Aristotle

A key distinction that separates man from our primate ancestors is the development of a higher language. It is hard to say which came first – the ability to make complex sounds using our larynx, or the ability of the brain to analyze and articulate the complex sounds. As a result of this happy combination, a wondrous thing happened – humans began to be able to carry on communications using a set of abstractions called words. Once language was invented, the economies of scale changed forever in the animal world.

To understand how we were before the development of language, we need to study the grouping behavior of our nearest cousins – the great apes. Gorillas have been observed in the wild living in groups of up to about 50 individuals - under the watchful eye of the dominant male. Chimpanzees and other great apes have also been observed to live in groups – but the numbers in each pack rarely goes past the very low hundreds. Groups that become too big tend to split up and go their separate ways.

There is a reason for this, and it has to do with how many individuals can be reasonably organized into a dominance hierarchy without the use of a higher language. These dominance hierarchies fall apart very quickly if the leader is not in direct communication with any part of his or her following for an extended length of time.

Language changed all that. There has always been strength in numbers, but now small tribes could create a common understanding and band together into larger tribes. If the occasion arose, more than one large tribe could link their forces together to control a region and its available resources. The tribes that were better at mustering their resources in battle were able to conquer other tribes and subjugate them. Over time, the language and customs of a few dominant tribes in a region became accepted widely and began to define the ethnic and cultural identity of that locale. Governance suddenly became a big issue.

Of Tribes and Nations

The biggest advantage of banding together was to give the tribal unit a better chance at survival in a hostile environment. Once the tribal unit had sufficient critical mass to pose a serious threat to any would be attackers, and to bring down big game – humans naturally found themselves in a situation where the mere act of survival was much less than a full time job. In other words, our ancestors suddenly found that they had time in their hands for erstwhile frivolous activities like arts and entertainment.

What started as frivolous activities, over time became an integral part of culture and civilization. No longer was it possible for one person to be an expert in all aspects of human knowledge. Specialization began to develop. Let's take an example from the field of medicine. First there was the tribal medicine man who would use physical concoctions (herbs, roots, etc.) as well as the metaphysical (chants, rituals, etc.) to look after the well being of his or her people. Then came the physician who would look after only the physical ailments that people were afflicted with. Then came the specialization of the physicians into more specific fields – like Gynecology and Dentistry. Today, even Dentists come in many stripes, e.g. Oral surgeons, Orthodontists, and so on.

One of the specializations that soon emerged was the one of governance. Several related skills began to be grouped together to define the ruling class – which in India came to be known as the Kshatriyas. The art of making war was probably skill #1. But it was not sufficient to just conquer – one had to maintain order

and create prosperity in the region under control. The system of laws and justice was invented. So was a system of taxation and concentration of wealth amongst the rulers.

This was the beginning of the feudal system – which has worked remarkably well throughout most of our recorded history. It boils down to a very simple concept, ALL responsibility (protection, taxation, justice, etc.) of a given population resided in the hands of the ruler. So, while there was specialization in the study of the various aspects of governance, there was little physical separation of who got to carry out the responsibilities and privileges.

To govern larger regions, a hierarchy of rulers began to form. At the very top was the Emperor, who ruled over several Kings, who in turn ruled over the Dukes, Earls, Barons etc. The feudal contract was really quite simple between the Lord and his vassal. In return for 'protection,' i.e. the privilege of keeping one's own position – the vassal would supply the Lord with a certain amount of 'loyalty,' usually measured in cash and other resources channeled upwards on a regular basis.

It is the simplicity of this contract that was also the feudal system's biggest shortcoming. It did not make for a unified society. Not only were the rulers classified according to their status and influence, but all of society was stratified along the lines of inherited social roles and responsibilities. Nowhere was it more codified than in Ancient India, where the four main classes came to be known as Ksatriya (rulers, warriors), Brahmins (priests), Vaishya (traders, farmers) and Sudras (serfs, menial laborers). This system of division was known as Varna – meaning Color – or the Caste system. It had its roots in pre-history when the Aryans (who were fair skinned) conquered the local dark skinned people (the Dravidians) and established themselves as the upper classes – primarily Ksatriyas and Brahmins.

Feudalism - The Price of Protection

Even amongst the Ksatriyas, the mid level ruler cared more for the balance of payments (i.e. the price of protection), than specifically which Emperor or even which dynasty was occupying the throne

at the highest levels. At these exalted levels, quite often there were bloody wars of succession, and the ultimate winner would often kill off not only their direct competition but all their family - and anybody else who might have a strong affiliation with the losing faction. It was much better for the mid level vassal not to get too involved with the rulers at the very top. This philosophy extended to situations where the rulers were from a different religion, or even a different nationality.

So powerful was the Feudal Contract in India that an external invader could come in with relatively small forces and knock off the guy at the top – and assume the feudal contract of all that were under them. In spite of a thousand years of rule by Muslim Emperors, most of India remained Hindu, and the ranks of the mid to low tier vassals was filled with Hindu Ksatriyas. Even when the British came to India, under Robert Clive in Bengal in 1757, they were able to knock off the regional ruler, Shiraj-ud-daulah, with a paltry force of about 3,000 men and the connivance of the General of the Army (Mir Jafar) who accepted British vassal-hood. Empire builders from Alexander to Chengis Khan have all used this weakness of the feudal system to their benefit.

Feudal Governance and its applicability today

The feudal system (or a variation thereof) is arguably the most prevalent form of government as of this writing (2006). With the firm roots that democracy has taken in many countries, this may sound like an outdated statement. Let us use a measuring stick to determine how much truth there is to this statement.

Of the 6+ Billion people inhabiting planet earth in the early 21st century – only about 1 Billion can be counted to be living in countries we would consider developed. This would include the USA and Canada from North America; Japan and Singapore from Asia; the 25 countries within the European Union; and Australia and New Zealand from Oceania. Even if we include countries like Russia, and Israel – the population within the developed world does not much exceed one billion.[1] Democracy is not the universal method of governance even in these countries – but it is close enough for us to categorize them as belonging in the democratic camp.

Chapter 3

The traditional kingdoms and principalities are easy to understand as feudal aristocracies, but what about areas like China and the Indian sub-continent where close to a half of the world's population lives? For close to 40 years after independence, India has essentially had one ruling family at the top – three generations of the Nehru-Gandhi family. It is only recently that there has been a viable alternative government – and even then it was a somewhat shaky coalition. The other four countries in the sub-continent - Bangladesh, Nepal, Sri Lanka and Pakistan – still have a long way to go towards a functional and inclusive democracy. The thinking process in circles of government continues to be feudal – i.e. protection for loyalty. The bureaucratic process is still such that it still takes a lot of loyalty (monetary contribution, legal or illegal), to get things going. However, things are slowly changing with more transparency and rule of law in Government affairs, but even with that we can expect that the feudal system will continue to be dominant till about the quarter century mark (2025).

Changes are happening in China as well, with a new generation of leaders having taken over. China's top leadership has been marked by long periods of power for the governing individuals, not unlike the erstwhile Soviet Union or the extreme case – Cuba. The current set of changes in China appears to be leading to more economic, if not political, self determination. At its core, though the communist system of government sticks to the tried and tested Feudalism/Theocratic combination – with two significant variations. The first is the elimination of a formal class structure. The second is the substitution of the Communist doctrine for religion – which we will discuss further in the Theocracy section.

Another sign that feudalism is running in strong undercurrents is the presence of pervasive corruption at many levels. Unhappily, this is true of most of the developing countries today. Bribery sounds like a terrible practice, but the concept is essentially at the core of feudalism. The person in a position of authority extends some protection or patronage in return for some loyalty consideration – whether it is cash or kind. Such an arrangement is also at the root of organized crime – and we know that even the developed countries have their own share of this! As we will

see later, influence peddling is not all bad – and a system could be found by which the benefits of such peddling accrues not just to the individual in power – but to the society at large, and with a large degree of consensus on the happening trade.

In the gradual transition from a feudal system to democracy, we must note that a lot of constitutionally democratic states may not be functionally democratic. Failure may happen at many levels, but the symptoms are often quite clear. One indication is extended rule by a single individual or family. Another is sustained corruption within the political process. Another may be that some constituents feel totally disconnected with their political rights. Without delving into statistics, let us note the considered opinion of the Aga Khan, noted philanthropist and the spiritual leader of the Shia Ismaili sect.[2]

> *"I estimate that some 40% of the states of the United Nations are failed democracies. Depending upon the definitions applied, between 450 million and 900 million people currently live in countries under severe or moderate stress as a result of these failures."*
>
> *- Aga Khan, April 2005*

Even if we fast forward 20 years, and assume that a well managed set of transitions actually happen in the Indian sub continent, China, and other places where democracy is slowly taking root – many aspects of the Feudal system would continue to live within an otherwise majority based democratic system. The Feudal system of command and control still has a lot of applicability in modern day institutions ranging from the military to corporate hierarchies. In most situations these institutions work within a strong civilian framework and is constituted within a set of common objectives - providing a counter balance to budding autocrats. However, even in the military engagements of the early twenty first century – such feudal behavior is sometimes quite apparent. During the war in Afghanistan, it was common for rival Afghan commanders to be in touch with each other using short range radio – and for them to even negotiate for the individual fighting units to change affiliations depending on which side appeared to be winning! There are other human institutions that

continue with the age old feudal traditions – even within the most progressive of countries. Gangs and narcotics traffickers organize along lines of loyalty, as do most forms of organized crime. It still is the organization of choice when a group of humans are thrown together and survival is on the line.

In net, it is too early to say that democracy, as a form of governance, has developed a majority following within our human civilization. Even when the intentions are there, the execution has often faltered – and the feudal system is still very much the incumbent system of governance in the early twenty first century! Even where Democracy is firmly entrenched – we will see there are plenty of advances still to be made towards more inclusive and responsive governance.

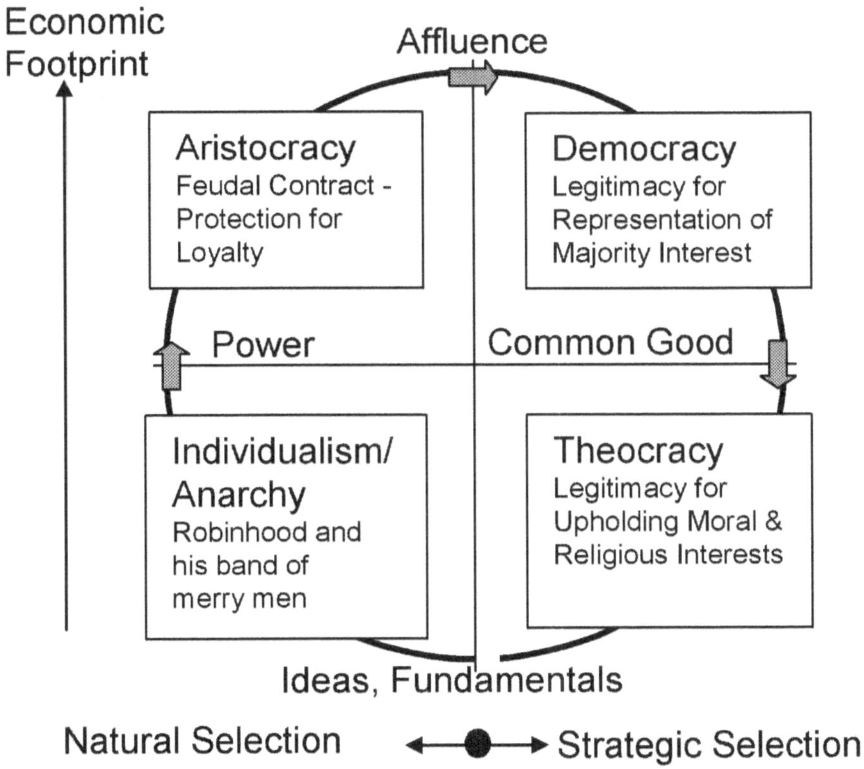

Economic Footprint

Affluence

Aristocracy
Feudal Contract -
Protection for
Loyalty

Democracy
Legitimacy for
Representation of
Majority Interest

Power | Common Good

**Individualism/
Anarchy**
Robinhood and
his band of
merry men

Theocracy
Legitimacy for
Upholding Moral &
Religious Interests

Ideas, Fundamentals

Natural Selection ◄——●——► Strategic Selection

Figure 3.1: Mapping the main forms of government onto the Mahashrama cycle

<u>Democracy and the Common Good</u>

As we have observed, human beings have lived most of our history (and pre-history) under the feudal system. Although democracy as a concept was first made semi-practical in ancient Greece, at that time they still practiced slavery – and not everybody had a vote. Real democracy did not become possible till the last hundred years or so, where most progressive nations gave up slavery and awarded full voting rights to women and minorities. In its most basic essence, democracy adds one more equation to the Feudal Contract – which in its original form now becomes less relevant. The new contract is the Legitimacy Contract – which is between the majority of the population and the elected leaders. In its most basic form it says – If the rulers continue to represent the Majority of the interests of the Majority of the people– they will continue to receive the support (or legitimacy) for their rule.

Just like the Feudal system had an Achilles' heel, the democratic system has one as well. This too has to do with the "war of succession" – in this case at the ballot box. The system flounders if the Majority of the people doesn't care to (or are unable to) find out if the rulers are actually representing their most critical interests. It depends on a reasonably well educated and informed electorate with a keen interest in the electoral process. Case to point is the large number of countries who have experimented with democracy – only to fall back into the tried and true feudal system (e.g. Ancient Rome was a republic, and then came Caesar).

So where does the great experiment in Governance called Communism fall in all this? In spite of all the ideology, Communism (in its mature state) is much closer to the Feudal system than democracy. Like the different levels of rulers in the feudal system, Communist party leaders essentially control almost all aspects of the life of the people in their jurisdiction. The legitimacy contract is really quite weak, as the same party decides what happens – often leading to very entrenched leaderships like Chairman Mao and Joseph Stalin. There is also a strong undercurrent of Ideology that comes very close to the Theocracy that we will discuss next.

The rise of Theocracy

Theocracies do not need a Legitimacy contract – as their
legitimacy is assumed to come from God. Replacing the
Legitimacy contract is the Moral Imperative. Throughout
history powerful rulers have enlisted the help of priests to paint
themselves as 'divine' or 'anointed by the gods'. Theocracy is an
attempt to separate out the person of the ruler from the divine
aspirations, and substitute instead a rather narrow interpretation
of the relevant scriptures as the guiding principles. Theocracies
can often exist in parallel with other form of governments like
Monarchy (as it did through most of history) or even Democracy.
Theocracies can often be recognized by how the country defines
itself – say, in the Declaration of Independence or Constitution.

If we take a look at figure 3.1, it might look like Theocracies and
Aristocracies are far removed from each other. Why then is it
so easy for a Theocracy to become an Aristocracy? There is
actually an in-between phase that passes very quickly – called
Individualism or Anarchy. Typically, there will be a power struggle,
and all remaining contenders will be vanquished as a select
few consolidate their hold on power. One only has to look at
prime examples like Joseph Stalin and the gang of four in China
to understand that no amount of ideology can take the basic
competitive instincts out of man.

When an individual or group consolidates power, it does not
mean that there is no need now for the grass roots theocratic
organization that put them in a position of power in the first place.
In most cases some kind of coexistence arrangement is worked
out by which the Theocrats rule supreme over certain aspects of
social order, while the rulers go about their own princely duties.
In ancient India, the Brahmins were entrusted with everything
Spiritual – but were not allowed to be landowners or rulers. In the
twentieth century the theocrats in Saudi Arabia (of the Wahhabi
sect) were entrusted with the education responsibilities for
the youth in the kingdom. This kind of affinity across opposing
quadrants are seen elsewhere – which we will examine in a
following section.

Throughout recent history Theocracies have been a powerful force in unifying a people - and have them condone or commit incredible acts of brutality in the name of religion (e.g. the Balkan Wars). They have also been much more capable than plain old stratified Feudal Aristocracies to motivate the common person to undertake equally incredible acts of sacrifice.

As noted, over the longer term, the steady state for a theocracy ends up closely resembling the feudal system. When the grass-roots fervor dies out, the people that are positioned at the top will consolidate their powers and rule the country not much unlike the Caliphate of 1000 years ago, or the princely Sheiks of the Middle East today.

This is true of Theocracies that are not based on religion as well – e.g. Communism. The belief system may be a system of desired behaviors for social development that are widely held – like the ideological fervor that gripped the masses during the early stages of the socialistic revolutions in Russia and China. The end result appears to be much the same, however, as a few individuals in position of power will consolidate their domain so completely that it begins to look much like the Feudal system of old.

Theocracies in Conflict

For Theocracies to thrive, the best conditions are exactly the opposite of the free flow of information and broad based education that are so critical to democratic functioning. To site an example from Iran – where a Theocracy (as of this writing) still holds on to power and controls vital social apparatus like the judiciary. There we hear about laws that prohibit "disturbing the Public Mind and Insulting Sanctities" and journalists having to go to jail for their web logging activities. To preserve order, Theocracies needs to preserve the sanctity of their beliefs – hence the rivalry between Democracy and Theocracy is often played out in the public information domain.

In the conflict between Theocracies and Democracies, this same free flow of information that is the economic and political lifeblood of a secular democracy can be used to bring harm to the democracy as well. Let's take the example of the September

11 (2001) attacks. The attacks were so well planned because all the information was publicly available– from flight schedules to airplane types to even the dimensions, location and structural details of the targets. The perpetrators even had access to flight schools where they could learn to fly without having to learn how to take off and land! Since the aforementioned attacks, the average traveler has had to tolerate longer lines at airports and heightened security costs, and the whole airlines industry seemed to have gone into a tail spin (especially in the US). Unfortunately, this is often the cost that democratic societies pay whenever conflict arises with a motivated and well funded theocracy.

By placing certain Moral Imperatives at a higher plane than the good of the individual or even the good of the majority of individuals, tremendous emphasis is put on the 'prophets' or the interpreters of these moral imperatives. In a secular democracy the interests of a majority and minority are usually not so far apart that the majority will be able to go very far just by exploiting the ethnic or economic minorities. On the other hand, a theocracy that rigidly defines it adherents, and specifically excludes the minorities or 'non-believers', can easily bring the wrath of the state machinery on them. The persecution often does not stop with the living! It might even be perpetrated against the historical vestiges of their long-lost existence (e.g. the Bamiyan Buddhas in Taliban Afghanistan).

In historical terms, Theocracies, like Communist Regimes – are an endangered species. With globalization, no part of the world today can be isolated effectively from the Radio, the TV and most alarmingly - the Internet. Even with full control over the curriculum of school age children, the Theocrats cannot forever isolate their youth from the alternative ways of thinking that exist elsewhere. An even more insidious problem for Theocrats is the problem they encounter when prosperity strikes the community. With prosperity comes the allure of forbidden goods (e.g. alcohol, outlets for libido) that quickly leads to a parallel society where the privileged live by a moral conduct that is positively blasphemous! Under attack from both the inside and the outside, it is easy to see why Theocracies turn so often to militancy and brainwashing of children as their method for survival.

So what is the Theocracy to do when faced with such challenges? Do they ride quietly into the sunset or do they define a last frontier that they are willing to defend at all costs? A lot depends on the ideologies contained in the Theocracies as to whether fight or flight is the preferred initial reaction. Islam, for instance, is a religion that was tremendously successful in combining the benefits of a peaceful (and resource efficient) internal society with a relentless pressure on external entities (states as well as individuals that were not adherents of Islam) to convert by choice or by force. The same can be said of Communism or to some extent Capitalism (as an Ideology) in the heyday of the Cold War.

The price to pay, of course, was a very firm doctrine (individual austerities as well as laws of society) that could be applied to almost any non-believer or even an occasional believer – to show why they weren't adherents of the one true faith. Hence, more and more, the relentless external pressure tended to become inwardly directed, especially as there were no easy conquests to be had of new adherents at the peripheries of the theocratic ideology. A few strategic defenders of Islam determined that the real threat was the burgeoning waves of democracy and global markets – and decided that the real enemy was the perceived leader of the democratic world – the USA. Hence we have the series of attacks on US interests, culminating on the most horrific one – on 9/11/2001.

If Fight or Flight are the only options – there is no end in sight to endless bloodshed in the name of preserving the sanctity and purity of a Theocracy! The enemy then is all around, and there is no place to escape to. Is there a third option that could lead to a peaceful coexistence without mankind becoming polarized between the perceived axis of good and the axis of evil? Do the labels of 'good' and 'evil' even make sense as an absolute metric when it comes to a clash of cultures?

Baby Steps towards Conflict Resolution

We need to refer back to the Mahashrama cycle to figure out an end to this conundrum. Four levels of understanding are needed to really solve a problem of this magnitude. The first, stemming from the Brahmacharya stage – is the education of the new generation to have a profound understanding of the whole of reality (Brahma) – and not just the doctrine of the forefathers. The younger generation needs to be equipped with all the skills – material as well as spiritual to be as successful as they can be. Breeding desperate people that cannot compete in the greater economic reality only serves to prolong the agony.

The second requirement is borrowed from the Grihastha stage – the need to provide economic growth opportunities so that people can enjoy the fruits of their labor. Without a good Grihastha stage it is very difficult to have a good Vanaprastha stage – hence it is important to build up the economic base of the population. Hence, even before we get to matters of ideology – grass roots economic productivity and growth needs to happen.

By the time we get to the Vanaprastha stage one might argue that a generation has passed by, and we have not yet addressed the root conflict of ideology. We have already seen how economic solvency seems to encourage the creation of a class of people with profligate lifestyles. So there are now more people in the enemy camp – so to speak?

The answer to this question is very simple. Except in a democratic environment - there is very little value to numerical superiority today. There is benefit to economic and technological superiority. Any society that wants to remain competitive long term must build their skills and economic base. An agreement with a vastly weaker opponent can be broken at any time with relative impunity – especially if there is no enforcer around. The minimum that the Theocracy under attack needs to do to be able to survive long term is: embrace all-round education and skill building, and also provide its adherents with the means to enjoy the fruits of their economic development. As we will note in Chapter 4, without a successful Brahmacharya and Grihastha stage, the population cannot be expected to successfully enter the Vanaprastha stage where the social benefit to the world at large can be demonstrated.

The opponents of a long lasting peaceful coexistence (from either camp) might do their best to derail both the skill-building and economic efforts. Those with a vested interest in the status quo (e.g. the proverbial arms dealers that are supplying both sides) may try to subvert the process of establishing a functioning Brahmacharya and Grihastha stage in an embattled population. Yet, this two-step developmental approach, if internalized by the majority of the people, can prove surprisingly robust to subterfuge. The reason here is quite simple; this strategy does not depend on a charismatic leader or a feudal hierarchy. Death or destruction brought about on one part of the population can be repaired elsewhere – and sooner or later there will emerge the Vanaprastha leadership that can rise above the historical differences and employ a 'Win-Win' strategy in the solution of the long standing problem.

The biggest contribution from the Vanaprastha stage is the emphasis on the common good of both the parties in the conflict. This is not the situation where one ethnic group or religion is persecuted or economically exploited by the other. This is where the common aspirations of all parties can be laid on the table and a plan drawn up to maximize the overall benefit to be gained by all parties. If one truly and wholeheartedly goes through the exercise, it is very likely that the answer will not be that the combatants are isolated, each given a domain of play – and steps taken to keep them out of each other's hair. Again, the overall vision of how the whole is greater than the sum of the parts needs to be widely accepted by both sides to show that cooperation is a lot better than strangulation. In Chapter 4 we will further develop the concept of Yogic Empathy to describe this process.

In the Sanyas stage we look to simplify the essence of what it is to be a practitioner down to the very basics. The strategy here is the opposite of the doctrine of MAD (Mutually Assured Destruction from the Cold War era) – which we will call MAP or Mutually Assured Preservation. We will cover more details of MAP also in Chapter 4 – so this will be just a brief mention. The first part of MAP is the capture of all the information about what it means to be a 'practitioner' in a good amount of detail, along with the art, artifacts and even the specific ways of thinking that define a

particular group or ethnicity. We could call this the Cultural DNA of the group. In its idealized state, the saved data is sufficient to allow a re-enactment, as it allows a future student to virtually enter the life (or even the person) of a practitioner. Converted to bits and bytes and artifacts, the essence of what it is to be – say – a Navajo Shaman would live on forever. At a high level, the first step of MAP is to lay out what it means to be one of the practitioners of the ethnic group, so that each party knows what it is that the other stands for and is looking to preserve. Hence, this process is actually independent of who the antagonists may be in the conflict.

The second part of MAP is to create a composite of the interaction of the lifestyles and cultures between the two adversaries. Almost always, there will be some parts of the perspectives of one that will complement the points of view of the other —and the purpose is to create a composite that is greater than the sum of the parts. In the case of the religious interaction between Islam and Hinduism in India – much of the story would be lost if we did not cover the Sufi Saints who preached a religion that was an amalgamation of the two, or the rise of Sikhism. We would also lose the perspectives of the great unifying Emperors of India like Akbar and their efforts to bring out the best from multiple religions. In the conflicts between two points of view, it is often easy to lose sight of how the world benefited because this cross pollination actually happened. Just the process of capturing and sharing the Cultural DNA of both conflicting entities/groups would itself go a long way towards creating a common understanding of the human condition from both sides

Individualism and Anarchy

With the advent of personalization technology (especially over the internet), the most basic form of organization and governance is set for a comeback. In spite of all our social conditioning, the human animal is, after all, an individual with his or her unique set of hopes and aspirations. In the olden times it was extremely difficult for the average Joe (or Jane) to leave his or her imprints on society – but that is changing rapidly. Already, in many countries (including the USA) the voters vote not only on their representatives, but also on issues which are proposed from the general population.

What is deficient in the current voting system is that the individual opinion counts only at discrete times (typically once in 1-5 years), and the voting is along the lines of pre-fabricated scenarios. With the right access to technology, there is no reason that an involved electorate would not want its opinion heard every month – or every week – or even every day if the issues were pressing enough! Some semblance of this can be seen already with the results obtained from opinion polls. With the introduction of technology, there is no reason that for sufficiently weighty issues opinion polls cannot become real polls. One more technological leap forward - and individuals may not only be able to have their votes counted real-time, but also have a participative interest in formulating the issues of the day.

Moving up the organization ladder from Individuals, small groups of individuals may desire to create their own identity. Individuals may participate in one or more such groups – which may even be at cross purposes with each other on certain issues. Is it even possible to hear all the voices of the individuals and the groups, and still be able to synthesize enough intelligence on a real-time basis - to be able to formulate an effective course of action?

To some extent, such a process is happening on a limited scale in governments like the U.S. government. There are lobbyists for all kinds of vested interests, including industry, labor, environmental groups, and so on. These people are trained to exert more influence than just the voting power of the constituents that they represent. On the surface, it may look like the subversion of the democratic process – with political contributions taking the place of bribery and corruption in other governmental systems.

Upon deeper inspection, it becomes clearer that there is a much bigger unfulfilled need for individuals and groups to be heard, and that the majority based democratic system has to be extended to address this need. Today it is the people in position to throw money or influence at a target outcome, that hire lobbyists in Washington (or other nation's capitals) to do their bidding – but tomorrow's governmental systems could formalize the process. It is not necessary that under-represented groups should need to pay for governmental favors. But if they did, the trick will

be to figure out a system for turning any economic benefits of such an activity over – not to the lobbyists and politicians – but to the government's exchequer. An example today is how electromagnetic spectrum is made available for wireless cellular services. It is auctioned out by the government in some countries, while in other countries it is just 'given away'.

To understand why it is becoming more and more critical for us to hear every single voice crying to be heard – and not just listening to the voice of the majority once every few years - we need to recognize two trends in modern society. The first trend is the gathering beat rate and ever faster pace of happenings in this world. The second is the fact that all individuals and nations are getting more intertwined in their daily operations - to such an extent that nothing is truly local any more. In combination, we have the scenario now that "no Man is an Island" - and we have to look at problems as if they are happening to the body of our Composite Human Civilization.

If we cry out loud, and receive no help – after sufficient tries it is human nature to do one of two things. The doves will accept it as the way things are, and stop complaining. The hawks will decide that this is a cause worth fighting for or even worth dying for.[3] Even if a very small set of individuals think like this, this could spell disaster in today's connected environment. One desperate person or group could potentially take down the whole ship with them. In the post 9/11/01 era, that's an uncertainty all humans have to face up to regardless of where they live in this world. Again, the tools and the attitude to reach out with are within reach – as we will discuss in Chapter 4.

The End of the Fortress approach to self preservation …

Those days are behind us when a ruler could purchase safety by surrounding themself with trusted people, and by employing a small army of food tasters to taste (and verify as safe), all that they ate. Today, no individual or group is truly safe if even a small group of well funded and highly motivated miscreants decide to take them out. It may look like just so much collateral damage if thousands of other individuals, including the miscreants themselves perish in the ensuing inferno.

It is an undeniable aspect of military technology, that for every two steps it takes forward in offensive capability, it only takes maybe a single step forward in defensive capability. Even 50 year old technology in the wrong hands can paralyze the world. We are not arguing that we ought not to protect our offensive weapons from falling into the wrong hands. What we need to recognize is that, even with our best efforts, technology cannot be forever held in friendly hands.

Sooner or later humans will be faced with deployable nuclear weapons in the hands of a party or parties that do not care if they are vanquished from the face of the earth – as long as they get to make their point! No amount of fortress building will spare us then. Before we get there, we need to permeate the new philosophy that we have referred to as Mutually Assured Preservation (MAP) – which depends not only on the preservation of our various lives and lifestyles – but at some level the very 'purpose of life' that drives us to do the things we do. Technology has a key role to play here – and openness to an all-inclusive form of government is another key enabler.

… And the Birth of Inclusive Governance

So how does this evolving change in our security interests effect the type of government we choose? In this section we are talking about Individualism and Anarchy – which till now has signified a lack of government. Yet, in it is the core idea of an inclusive system of government that does not let any of its constituents or their ideas down without due consideration. This is democracy with a twist. In a democracy, the majority rules – and the opinions of the majority count. In an Individualistic-Democracy the thresholds of inclusiveness and perspective are brought down to the level of individuals and small groups. Even apart from enhancing the obvious safety and security issues, this also has the benefit of utilizing many more of the world's 6 billion minds in figuring out the best future that there is to be had for all mankind. And yes, technology – especially internet based technology with its extensive reach - will help make this vision a reality.

The Axis of Good vs. The Axis of Evil

We have already noted that there is a strong affinity between the Theocratic and Feudal forms of Government, and how this has been the number one combination, from before the times of the Pharaohs, in maintaining a strong focus of power in the hands of the ruler(s). We can surmise that an equally strong (if not stronger) bond exists between the Democratic form of government and the concept of Individualism/Anarchy. Is it any coincidence, then, that these types of pairings tend to form along segments that are on opposite sides of the Mahashrama cycle?

If we connect the opposing segments using a line – we end up with an axis or worldview around which the society can be thought to be structured. We have noted earlier that the main system of government for humans still is (and will remain for some time to come) the feudal system or its variants (like communism). There is no doubt, however, that Individualism and Democracy are on the rise, and that the old guard naturally feels threatened.

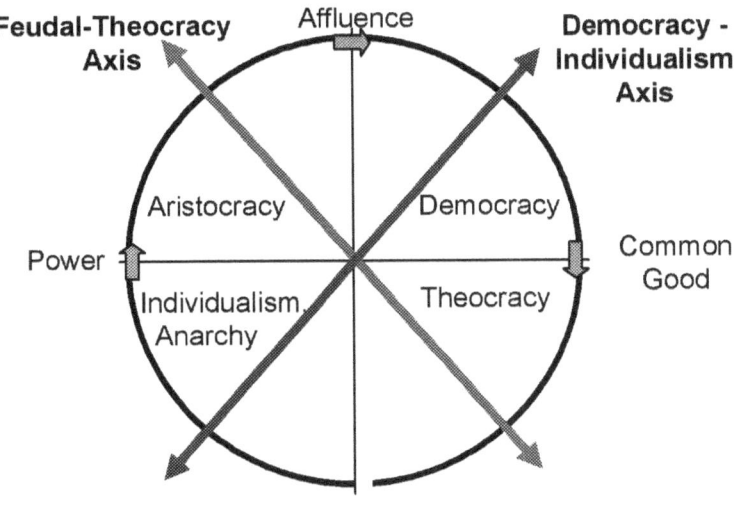

Figure 3.2: The Feudal-Theocratic Axis vs. the Democracy-Individualism Axis

Figure 3.2 shows why the systems of government appear so incompatible to the other camp, that phrases like the "Axis of Evil' are used by each side to describe the other. For the first

time in human history we are looking at the prospect of most of humanity moving out of the Feudalism-Theocracy axis over to the Democracy-Individualism axis. Even amongst the democracies in the world, there is considerable leeway in interpreting which part of the quadrant the axis runs through.

Let us take the two party political system within the USA as an example. The conservative Republicans might see the axis very close to perpendicular, with a strong emphasis on economic prosperity, and a few, relatively clean governing principles. The more social Democrats may see it as very close to the horizontal axis, trying to represent more of the common good, and extending coverage to include more diversity both in terms of ethnicity and ideas. This is illustrated in Figure 3.3. Economic interests (like big corporations) will typically favor the axis that is close to vertical. In times of severe social and economic stress, people might favor an axis very close to vertical as well. When times are less stressful economically and socially, it is expected that the higher degree of common good and diversity embodied in the axis close to horizontal would win out. These are broad generalizations, however, and many other factors (including the quality of the leaders) would play out in an actual election.

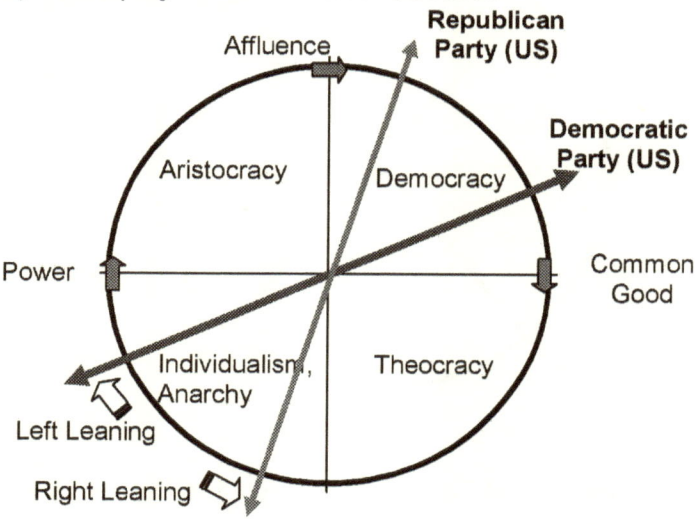

Figure 3.3: Left and Right leanings within a democratic system

Reality Check – Dealing with Conflict

Certain experiences, even if done under the context of an otherwise democratic government, can cause tremendous conflict within the people that are asked to undertake such a task. Let us take the example of the draftees who were trained and sent to Vietnam to fight the war against Communism. Exposed to a series of brutalities from either side, in which the poor helpless local population often suffered the worst of the fighting, it was natural to question whether the cause was worth the suffering.

Going back just a little earlier in history, it was more the norm rather than the exception that individual members of society associated with the armed forces perpetrated or endured tremendous brutalities, for causes that were little more than the wishes or even the whims of their rulers. These same rulers on the other hand generally legislated and maintained a decent civilized co-existence within their borders.

With the demise of the warrior class – and most other classes – how does the otherwise civilized human bring themselves to face up with the two massive realities of armed conflict? One, that for the collective entity to win, it is often necessary for the individual to sacrifice themselves. The second is the apparent hypocrisy of upholding a very high standard of civilized behavior within the country in normal peacetime, and a much lower – almost brutal and animalistic – standard under conditions that involve threat of war.

With a proper appreciation of the Mahashrama system, the same individual would be able to move their current mindset from one quadrant to the other - but needs to avoid getting stuck in any one segment for too long. We have previously touched on the concept of 'Dharma' which here translates to being the best person we can be given our current circumstances. Sometimes we do not choose our circumstance – it chooses us. When we are done being the best person we can be under trying circumstances, let us move on and create the environments around us that will truly bring out the best in ourselves and those around us.

Frozen in time:

The answer in the feudal days to any kind of opposition within the population was often to identify the leader or leaders and summarily execute or exile them. This was possible whenever there was a visible rebellion. There remained places that were remote enough and fragmented enough that even the most overpowering of emperors had a hard time bringing under their control. A case to point is Afghanistan – which resisted foreign occupation from the times of Alexander, and even faced down the British Empire at its peak. If one leader was killed, ten formed. Even the coming of Islam did not lead to unity, and the trend continues today.

Time appears to have stood still in such regions, with the organizing and normalizing influence of language still barely taking root. This observation is by and large true of any terrain where humans did not develop large contiguous settlements like towns and cities, but were geographically isolated in mountains and deserts and icebound landscapes. The great hope of technology is that it will bring such people closer to the overall human experience, where not only can they learn from the collective human accomplishments, but also contribute greatly in enriching it. Their lives are like a time capsule, and care must be taken in preserving not only the artifacts but also the overall human experience before their unique perspective gets lost forever. Another example of an ancient system that survives in a time capsule even today in many parts of rural India is the Varna or Caste system.

Varna, and its applicability today

The Varna, or Caste system, was practiced in ancient India, and essentially extended the system of classification (e.g. of rulers as in the Feudal System) to include the entire population. All of society was divided into four castes:

> Ksatriya – Warriors, Rulers
> Vaisya – Traders, Merchants
> Sudra – Servants, Laborers
> Brahmin – Priests, Teachers

Today, we have significant upward and lateral mobility in our societies, so obviously classification by birth or skin color is no longer as relevant. As one passes through the Mahashrama Cycle, however, we can note that the four Varnas are none other that a 'state of mind' or a Value system (Dharma) that one internalizes as they pass through the vertical and horizontal axes of the Mahashrama Cycle. The axis between the Brahmacharya and Grihastha segments is the Kshatrya axis – with a lot of emphasis put on competitive superiority. The Axis between the Grihastha and the Vanaprastha is the Vaisya or economic axis. The Vaisya were the economic class (traders and merchants) of ancient India. The axis going to the right between the Vanaprastha and Sanyas segments is the Sudra or service axis. The Sudras were the servants and manual laborers. And finally the axis pointing downward between the Sanyas and Brahmacharya – dealing with ideas and fundamentals – is the realm of the Brahmins. The Brahmins were the traditional priestly class, as well as the teachers. Since the core ideas of each classification are already contained in the Mahashrama cycle, we will not concern ourselves much with the Varna concept, but stick with the more universal principles of Mahashrama.

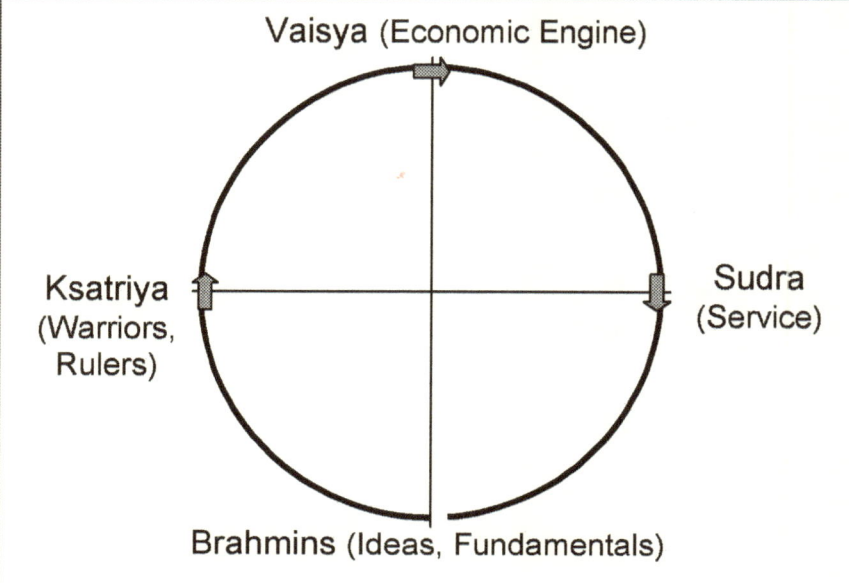

Figure 3.4: Mahashrama Mapping for the four Castes (Varna) from ancient India

To be very clear, however, the lack of social mobility associated with the Varna system is totally out of place in the world today – and needs to be considered as an artifact engineered by the ancients to maintain social order. Many parts of rural India still labor under the Caste system – but the barriers are slowly breaking down under the onslaught of social and economic mobility.

Forward in Time:

With all the changes happening in society, some of which goes to the very root meaning of being human – we must be forever willing to augment and extend our understanding of governance and what constitutes civilized society. Ancient building blocks that defined how society worked (like the Varna or Caste system of ancient India) need to be put in context, and not taken on its literal meaning. Even for relatively modern religions like Christianity and Islam, there is an ingrained exclusiveness that is conveyed upon the adherents as the 'Only true path to salvation'. This has convinced the ambitious and the powerful amongst us to take either their missionary zeal or the threat of the sword (sometimes both) upon the 'non believers' to bring them to the one true path. Such exclusiveness is just as obsolete as the exclusives implied by the caste system of ancient India. These distinctions between humans have served their purpose in preserving the social order or explosive growth that was engineered into them. Now is the time to recognize the full diversity that exists in our world, and to appreciate the diversity that we ourselves go through as we pass through the Mahashrama cycle.

Having considered what traditional baggage we need to leave behind – let us now consider what the essential building blocks ought to be for a diverse yet all-inclusive form of government based on Individualized Democracy. This is a key theme we will develop in Chapters 4 and 5, as perhaps the most critical contribution from the Vanaprastha segment, to help secure a successful and long lasting future for humanity.

Chapter 4
The Neo Vanaprastha – Achieving Segment Balance

Every moment of your life is infinitely creative and the universe is endlessly bountiful. Just put forth a clear enough request, and everything your heart desires must come to you.

- Mahatma Gandhi

Every person's needs are individual and unique, yet there are certain patterns of needs that emerge based on where someone is on the Mahashrama Cycle. These are the needs specific to the segment of the Mahashrama Cycle that, if left unfulfilled, would prevent a timely transition out of the current segment and on to the next. The vanaprasthi service leader would do well to familiarize themselves with these core needs – including the ones that they themselves would rely on to facilitate a healthy and productive Vanaprastha phase. As part of the Vanaprastha Network, the vanaprasthi can then work on helping each segment of our population develop to their fullest potential through the satisfaction of these needs - thereby preparing them to move on purposefully to face the opportunities in the next segment.

Abraham Maslow did some ground breaking work in the early part of the twentieth century to peg down what he determined were the tiers of human needs that are at the core of human existence.[1] Left unfulfilled, these needs hold back the individual from moving on to a higher level of human experience. These progressively higher needs were:

1. Physiological Needs – most basic – like food, shelter and clothing
2. Safety and Security
3. Belongingness and Love
4. Esteem (both self esteem and from others)
5. Self Actualization – the pursuit of the individual's true interests in life

In Figure 4.1 we see how the first four needs fit within the Brahmacharya and Grihastha segments – along with a few other needs that were not specifically categorized by Maslow. Self Actualization appears to be mostly a Vanaprastha need, with some spill over into Sanyas. Maslow had defined his needs as a pyramid, with the lowest levels having to be satisfied before one could graduate to the higher needs. With the more complete set of needs mapped on to the Mahashrama Cycle in Figure 4.1 - here too we will discuss the needs in sequence, starting with the Brahmacharya segment needs and working our way all the way across to Sanyas. It is important to note that the order of the specific needs within each segment is not as critical in the revised context. What is important in the Mahashrama context is the fulfillment of all of a particular segment's needs – which together serve as a measure of one's ability to graduate from one segment and move on to the next.

Brahmacharya Needs:

Physiological Needs

The Physiological needs are the most basic - food, shelter and clothing that any human would need to survive in the most elemental sense. Without these three the biological organism that we are at the physical level withers away and dies. Maslow recognized these as the most basic of all needs. These needs are especially keenly felt within the Brahmacharya segment because the developing humans are not equipped to provide for themselves. As we look at the newborn human baby – we immediately recognize that within the animal kingdom we humans are exceptionally helpless when we are newborn! Without the basic necessities of life being provided to us at an early age, it would be hard to even survive long enough to move on to the higher needs.

Physiological needs are associated with living a very basic healthy life – and as of this writing we are not even close to having this basic level of existence available for all our human brethren. Famine and disease are still ravaging vast regions of this world (especially Sub-Saharan Africa). Yet each person and situation is

slightly different – and the challenge for the neo-vanaprasthi will be to slowly and individually move a large section of humanity off of the brink of abject depravity. Closely related to the Physiological needs is the necessity to bring someone back from a Physiological disaster-zone by taking care of them when they are malnourished, sick or disabled.

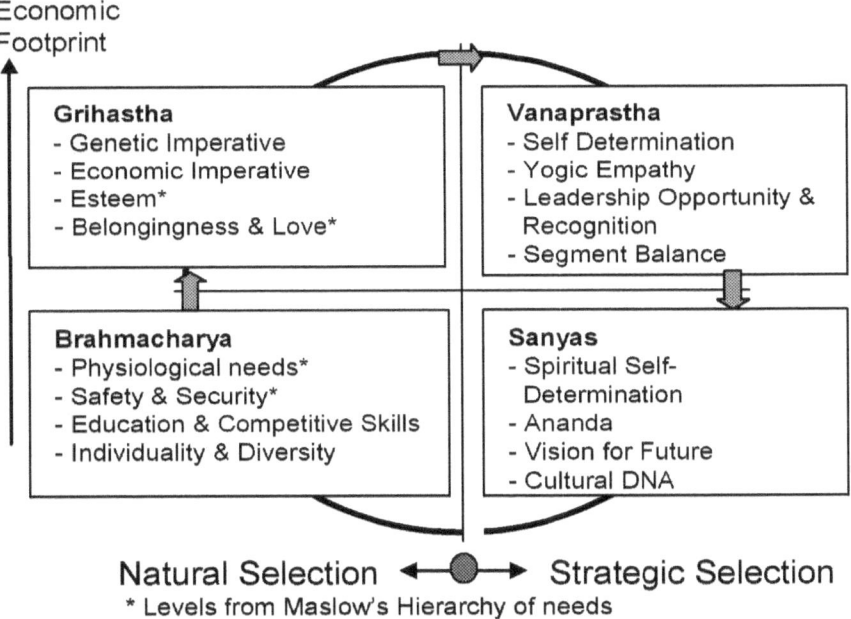

Figure 4.1: Opportunity space for the New Vanaprastha

Safety and Security

Safety and Security needs, according to Maslow's hierarchy, come right after Physiological needs. The shelter that protects us from the elements is a Physiological need, but also doubles as the protection for ourselves and our possessions from thieves,

vandals and the occasional hungry animal. Safety and security does not stop at the boundary of our homes – it continues to be a need for us as we roam within our neighborhood, or venture further out in pursuit of our education or livelihood. The safety and security needs for humans, individually and collectively, are such a critical element for future civilizational growth – that we take a deeper look at this topic alone in Appendix 1.

It is interesting to look at Information Security under the overall umbrella of security needs, especially as more and more of our existence moves over to an electronic medium (e.g. financial transactions). Going forward, it is expected that part of the motivation for us humans to really trust the electronic medium as an extension of ourselves - will have to do with how safely and securely we can extend our presence into this new medium. The trick will be to enable a high level of security without turning the process into a complex bureaucratic jungle.

Education and Competitive Skills

Once we have these two very fundamental needs taken care of, we can now move on to the core developmental need of Brahmacharya. The core need that comes from the very definition of Brahmacharya is the need to be educated, and proficient in the use of certain competitive skills that we will need to be successful later in life. This is the number one predictor of future personal success - and as such deserves recognition as an essential need for all humanity. Closely related to education is entertainment – which is when we are participating in an experience more for the aesthetics of the content than for its skill building ability. Entertainment cannot, by itself, be classified as a core need, but we must acknowledge that the entertainment process can be very effectively used to educate and communicate key messages.

Individuality and Diversity

The next need is for Individuality, as well as Individuality applied to groups (i.e. Diversity). This one may be a little harder to

understand as a core human need. Individuality relates to the
fact that as humans we are all individuals - and our perspectives,
needs and priorities are all somewhat different and changes
as we go through life. This is not to say that there aren't any
commonalities, but to establish that our core identity is very
individual - and then we can build on top of that as social
and ethnic groups, nationalities, etc. When we see children
(esp. teenagers) display rebellious behavior, and almost state
themselves as the antithesis of what their parents would like them
to be - we are often inclined to dismiss this a passing phase. If we
can accept that this is the time when Individuality is beginning to
express itself, it then becomes much easier for the well intentioned
vanaprasthi to relate to the unique needs of the developing
brahmachari.

Another reason to highlight the need for individuality is that
one's personal perspective can be quickly misplaced when
other individuals have to be selected (or elected) to represent
their views and opinions. The enlightened vanaprasthi looks
at Individualism as an asset and not as a threat. This is the
mechanism by which the individual viewpoint or needs gets
expressed and coordinated through the Vanaprastha Network.
Hence, it is vital that the vanaprasthi in the Vanaprastha Network
create an early, deep and enduring connection with the individuals
they serve – and try to relate to their core needs; as opposed to
the superficial needs they might sometimes put up as a ruse to
get attention. If we interpret the apparent need of children and
teenagers for personal attention as an expression of their own
individuality, then the vanaprasthi can use the satisfaction of this
need for mutual benefit - as they go about crafting a personal win-
win relationship with the impressionable brahmachari. If all goes
well, this personal guidance from the vanaprasthi would go a long
way to help develop the character and skill set of the brahmachari;
not only for the transition to the next segment (Grihastha) but for
their trip through the entire Mahashrama Cycle.

When the need for expressing their Individuality is applied to
groups – the same concept of self expression is now called
Diversity. The additional dimension that Diversity adds is the need
for free association between like minded individuals, or ones that

share the same history or ethnicity. Just as a means needs to be found to connect every human individual to the Vanaprastha Network - similarly the mechanism by which the Diverse associations of humans represent their interests also needs to be worked out. In going from individuals to diverse groups, some conflicts may arise between the individuals themselves and the groups that they are part of. These conflicts and their resolution are an integral part of the competitive Brahmacharya segment, and the vanaprasthi must be ready to step in with their enhanced conflict resolution skills when called upon to do so.

Grihastha Needs

Genetic Imperative

As we end the highly competitive Brahmacharya Segment with a high degree of self-sufficiency - we get ready to embark on the Grihastha segment. This transition typically happens with the selection of a mate and the accompanying preparation for the new-born children that are expected to follow. The 'Griha' in Grihastha stands for 'House' or 'Home', so the implication here is that we are setting up a stand-alone family unit. The Genetic imperative as part of the Grihastha cycle is not just an outlet for libido and sexual conquest - as it is sometimes perceived to be towards the end of the Brahmacharya segment. The individual grihasthi are now mature enough to plan out how to bring new human being(s) into this world and to give them the care that they deserve. This is a major step, imbued with great responsibility, and demanding of a high level of stability as we make space in our lives for one or more particularly helpless creatures – the newborn human baby! In most countries today, the average maternal age for the first childbirth is being pushed higher, as we try to line up enough resources (and courage) for a successful parenthood.

As of this writing we see birth rates steadily falling amongst the well educated in most countries, and some nations like Japan and Germany are trending towards a negative growth rate. On the other hand, strong population growth is still the norm in relatively undereducated areas, or where the religion encourages sharp population growth. This is a particularly intractable problem as

Chapter 4

we strive towards global equality for all humans – and something that the vanaprasthi must recognize if we want to prevent the world from getting polarized between the Minority 'Haves' and the Majority 'Have-Nots'. The population control experiments that have been tried so far have had mixed success in limiting birth rates, and have often resulted in a serious imbalance between the numbers of male and female children. The most significant experiment here is the Chinese 'One Child' policy – which though effective in limiting population growth has subsequently resulted in tens of millions more males looking for a mate than there are females. The root cause, in societies like China and even India, is the emphasis on having a male child – and the availability of selective abortion tools (or even infanticide as a method of last resort) to select the 'desired' offspring.

In most cases, the genetic imperative is so hardwired into the human psyche that the vanaprasthi has to do little to help the process along – except to pass along their blessings and best wishes to the newborn. The hard work for the grihasthi starts with the planning, and reaches a crescendo as they go about raising successful children through the timely fulfillment of their Brahmacharya needs. Speaking from personal experience, the grihasthi can definitely do with a lot of help in that department. This part of the work is very personal and family oriented, and can benefit from a Vanaprastha Network with deep roots that reach down to the level of the individual. The Vanaprastha challenge is different in the case of severe population imbalances, and macro level steps (e.g. governmental policy) will likely be necessary to protect society from their worst effects. Examples of these imbalances are: disparity between the sexes (as noted earlier in China and India), an overly high birthrate that threatens the supply of available resources, or even an overly low birthrate that threatens the continued youth and vigor of a population.

In any given population, it is probable that a certain percentage of people do not end up fulfilling their Genetic imperative. In fact, even today in many societies, it is common for young adults to give up their Genetic and Economic Imperatives and become a monk or celibate priest. In fact, even in the ancient Indian Ashrama system, a short cut was often indicated for the young

adults that wanted to go directly from Brahmacharya to Sanyas. In today's Mahashrama reality, it may be possible to forgo the genetic imperative – but it would be very difficult to create a shortcut through Grihastha by bypassing the other three Grihastha needs.

Economic Imperative

"The world doesn't owe you anything but an opportunity." So goes a quote form Charles Barkley – a colorful NBA Basketball player who used to play for the Phoenix Suns in the early 1990's. Charles grew up in a tough inner city neighborhood, where opportunities were few and hard to come by. Squandering these precious opportunities might have been more the norm than the exceptions in such situations – where the young adults were faced with numerous social plights like drugs, violence and high unemployment. A critical thing that the well meaning vanaprasthi must evaluate is the furtherance of the Economic Imperative through giving truly motivated individuals the start that they deserve. The Economic Imperative can be defined as the internal drive to work for a better economic future for our selves and our families, and the ability to preserve the fruits of our labor for future benefit. This is the motive power which makes the Market Economy we discussed in Chapters 1 and 2 work so well.

Many a heart warming story comes from rural Bangladesh where the Grameen Bank – with the help of a set of small loans, inspired the entrepreneurial spirit in people faced with an otherwise hopeless situation - and helped them start a successful small business. Similarly novel ideas come from remote parts of India (Kuppam, Andhra Pradesh) where one of the big names in technology – Hewlett Packard started a program to outfit entrepreneurial photographers with a back pack that became their portable photo-lab. This kind of approach goes against every instinct of big business and big government – how does one justify the overhead needed to support thousands of entrepreneurs in remote locations that yield only a few dollars per month in return?

The kind of outreach that is needed to support such a customer base can be expanded to include each and every person in a specific locality - but there needs to be a real caring vanaprasthi at the other end of the line to deal with the unique needs and opportunities that each individual faces. Much of the enabling technology already exists; as demonstrated by the web based personalization (e.g. My Yahoo) and support strategies (e.g. chat) that are becoming as a way of doing business on the internet. Can it be far away that a similar set of tools are used to reach every web-friendly human in the solution of their own problems - through guidance, consultation and fulfillment of a genuine need? The challenge for the neo vanaprasthi will then be: how do we make sure that the majority of the people, in a majority of locations around the world are not only able to meet the economic needs of themselves and their family - but also have enough left over to be able to move purposefully on to the Vanaprastha and Sanyas stages of life.

Esteem – both Self Esteem and Esteem from others

Esteem from others usually come from being able to express some qualities in our personal lives that society regards highly. It also comes from having a successful track record in life, whether at the professional level or in the family or community environment. As noted earlier, the need for esteem was recognized by Maslow as a key step in the hierarchy of needs. A vanaprasthi looking after the interests of the Grihastha (or even Brahmacharya) demographics would do well to build in a reward and recognition system into the overall process – such that extraordinary effort and superior results can be recognized and held out as role models for society.

Self-esteem, on the other hand, comes from within – and is usually tied to a sense of well being and accomplishment. Too often, self esteem is overly dependent on receiving esteem from others – and our own hidden accomplishments are swept under the rug. Appreciation of the good in ourselves goes hand in hand with the appreciation of the good in others – and is a key skill to learn at an early age for future success in life. Not all reward mechanisms that the Vanaprasthi sets up need to be

publicly acknowledged. An occasional private communication
on a individual's family or personal progress, coming from a
well respected vanaprasthi - can go a long ways to enhancing
an individual's self esteem. The giving of appreciation, as well
as the set up of a social environment that values giving and
receiving well founded positive feedback remains one of the key
responsibilities of the Vanaprastha Network.

We have not yet discussed the internal organization of the
vanaprasthi working in association with each other – which we
have referred to as the Vanaprastha Network. That a single
vanaprasthi cannot undertake to satisfy all the needs of a
large society is quite obvious – but how do the well intentioned
vanaprasthi go about getting themselves structured to become
the fail-safe safety net and organizing power behind all of society?
Can this be done with all volunteers, or do Governmental and
other institutions from the Vanaprastha segment need to get
involved? We will not be able to answer all the organizational
questions at this stage in the book. Let us start by recognizing
that the Vanaprastha Network has a very serious organizing
and directing role to play in our civilization – much as a brain
directs and organizes our activities as biological organisms.
For now, let us just say that the Vanaprastha Network depends
on the dedication of a large number of service leaders of the
Vanaprastha mindset. The key attribute of scalability that is
needed to be able to address the smallest needs of the individual,
and yet be able to scale up to the largest issues facing society
– will need a significant transfusion of new technology. We will
develop these technological requirements further in Chapter 5.

Belongingness and Love

The first circle of belonging that we usually have around us is
our family. Then we have our circle of friends, our neighborhood,
community etc. The social fabric that we build around us has
many benefits – including some of the safety and security needs
described earlier. The same social fabric can be very effective
in meeting the belongingness needs of the Grihastha (and to
a lesser extent the Brahmacharya). Love is a word with many

connotations – but in this context it represents the fellow feeling and caring that normally develops when a family or community group shares a strong sense of belonging.

Both Esteem and Belongingness needs are borrowed from Maslow's hierarchy. Together with the successful resolution of our genetic and economic imperatives, the satisfaction of these needs would likely satisfy the ambitions of the majority of humans as of this writing. The 'needs' that we will list for the Vanaprastha and Sanyas segment are probably less 'needs' in the traditional sense, but capabilities we will need for the successful conclusion of these subsequent stages. For the purpose of continuity – let us continue using the word 'need' – with the understanding that these attributes are required for us to successfully navigate the right half of the Mahashrama Cycle.

Vanaprastha Needs

In the process of figuring out how the Vanaprastha segment can help address the needs of the folks that are in the other stages of the Mahashrama cycle, it can be sometimes overlooked that the vanaprasthi have needs themselves that need to be addressed. This is for both their individual well being, as well as in support of their ability to benefit the other Mahashrama segments. A pragmatic vanaprasthi will accord these needs just as much serious consideration as the other needs we have discussed so far.

Self Determination

Self Determination is the process of figuring out who we are and what we stand for – and for finding the internal motivation to pursue the goals and activities that satisfy our soul (core identity and purpose). One of the sub-needs that comprise the overall need for self determination is the need to find our Relatedness – both in terms of identity and culture/philosophy. This may be expressed as a desire to research into our family tree and in discovering the kinfolk that we have spread around the globe. Another frequently expressed behavior is the formation of clubs and organizations to better express our membership in one or

more groups. To truly partake in Self Determination one also needs a high level of Autonomy (related to the Individuality/ Diversity need from Brahmacharya) and Competence (Education/ Competitive Skills – also from Brahmacharya).

> *To be self-determined is to endorse one's actions at the highest level of reflection. When self-determined, people experience a sense of freedom to do what is interesting, personally important, and vitalizing.*
> *- Edward Deci & Richard Ryan[2]*

Professors Deci and Ryan at the University of Rochester have developed a framework based on self determination, using which researchers across the world have structured coursework and training geared towards enhancing intrinsic (or internal) motivation. Intrinsic motivation is the opposite of 'extrinsic motivation' which works quite well in the left half of the Mahashrama Cycle. Extrinsic motivation occurs when an external reward structure motivates us to perform. In contrast, intrinsic motivation needs no external rewards, and the inspiration seems to come from a high level of internal reflection.

Just about all the needs we have discussed from the Brahmacharya and Grihastha segments have external manifestations that can be satisfied by a sufficiently involved Vanaprastha Network. The partial exception is the need for self-esteem which is also a key contributor to self determination. With self-determination we move from the realm of 'what' it takes to satisfy our needs to 'why' our needs exist in the first place. This is achieved through the 'highest level of reflection' – and is similar in nature to the deep introspection that often marked the beginning of the Vanaprastha stage in ancient India.

Yogic Empathy

> *It is Yoga that reveals the supreme secret: man is tomorrow's God, and God is today's man.*
> *- Sri Chinmoy (Indian Spiritual Leader and Author)*

What does Yogic Empathy mean? Let us start with the meaning of the Sanskrit word Yoga. Yoga, or Spiritual Union, is a practice from ancient India that allows the various levels of our existence to be combined into one during the Yogic unification process. Examples of Yoga include the meditations that allow our consciousness to mingle with our deeply embedded unconscious self. It also includes the series of physical exercises and the holding of postures that allows our mind to connect with our bodies. This is not a connection of conquest or ascendancy. It is the connection of being together as one. A core skill of Yoga is letting go of our Ego (even if temporarily) – and observing things as if from within a Natural Order. Judgments are set aside for the moment and a whole hearted attempt is made to identify with the core identity, aspirations and difficulties of the other. This is Yogic Empathy!

To those of us who have not practiced Yoga, we may be able to relate to moments of insight (or epiphany) where we get a glimpse of how our existence relates to the greater scheme of things. Those who are deeply religious submit their Ego to the will of God and can find tremendous inspiration in that manner. They also gain a perspective as to how they fit within a universal order. It is important to note, however, that religion is not a necessary condition for the practice of Yogic Empathy. It might even be a distraction – by raising barriers of who is deserving of an empathic connection in the eyes of a particular religion or sect.

The Yogic Empathy skill is the most powerful instrument that the vanaprasthi can use to bring about the changes we have discussed so far. To work well, this approach has to be tuned to the characteristic needs of individuals from the various stages of the human Mahashrama Cycle – including Sanyas. If a greater or parallel Intelligence is found in the Universe – this is also the skill that will allow us to come up with a mutually beneficial coexistence strategy.

Tools and Practical Aspects of Yogic Empathy

In a real world that is vastly complex, some tools will be needed to move Yogic Empathy out of the realm of the Spiritual to the Practical. For starters, parties seeking to establish an empathic

connection should have an idea of their own identity and aspirations - as well as the difficulties they face towards realizing those aspirations.

A practice that is quite prevalent as of this writing is Web Logging (or Blogging) where the participants seem to pour out their hearts, their stream of consciousness and their attitudes towards all things for other internet users to experience. It takes a conscious attempt to open up our lives to do that – and the exercise itself can help us sharpen our perception of who we are, how we think and what really motivates us. Bloggers today do not necessarily have to expose all their innermost and private (and possibly embarrassing) thoughts and actions – but are often surprisingly honest and transparent about their intentions and motivations. We hope this will continue to be true of people building an extensive web presence for themselves in the future. Even with the most intimate of bloggers, it is very likely that certain thoughts and ideas will continue to be locked away for special or restricted access – possibly to be shared with somebody they truly have reason to trust.

Organizing Information – from Web to Social Fabric

With millions, possibly billions of human web images on line, it will be an interesting challenge to match up the people with needs and the ones with the ability to fulfill those needs. It is expected that even relatively small needs can be identified and fulfilled quite readily – giving rise to a commercial market of grand scale and accelerated (if not instant) gratification. In fact it is easy to see that the same gratification network can be made to play for non commercial applications as well. Of primary importance to the vanaprasthi is keeping track of the needs of the whole gamut of constituents that they are seeking to serve. For the brahmachari it could be instant access to relevant information on areas that they are pursuing a mastery of, as well as the formation of virtual 'Buddy' groups with others of similar interests from across the world. For the sanyasi this might be where they build the temple to sanctify and immortalize their contributions to Humanity!

The transition of the Internet from a Web to a Fabric that binds us can be thought to happen when half or more of a given population has a permanent presence on this electronic medium. 50% is a rather arbitrary number – but begins to signify that the majority of the population now believes that they have a physical home as well as a virtual home within the web fabric. The concerned vanaprasthi also has to figure out how all these need fulfillment transactions happen in a safe, secure and timely manner. One of the key contributions here by the Vanaprastha Network would be to monitor the health of the emerging social fabric and determine that there is a high level of authenticity associated with the denizens (or storefronts) within this new web fabric.

Technically, there is no reason that Blogs of the future cannot contain all of the sensory attributes that we associate with our surroundings. Sometime in the not too distant future, we might be able to construct a full sensory (sight, sound, touch, smell and taste) virtual reality on the web, and literally be able to say "Welcome to my life" and mean it! The first step, then, for individuals and groups preparing for a Yogic connection is for us to internalize and be able to share:

1. Who we are, and the experiences that have shaped us
2. Our Aspirations, and
3. The Barriers we would like to overcome

The second part of the Yogic Empathy tools involves figuring out how the two parties involved can help each other out in a way that the whole is greater than the sum of the parts. This involves analyzing a lot of different scenarios, and picking a path that is most likely to succeed and be fulfilling to both parties. One of the tools used by companies trying to define a winning strategy is business simulations - where various scenarios involving the company, its customers, and the competition are simulated; and the various options sorted by risk and potential benefits. Similarly, the technically progressive vanaprasthi may want to work with the individuals they are helping, analyze multiple success scenarios and together pick the best options to pursue. Many of the most powerful computers in the early twenty-first century are used to simulate real world events – like climatic patterns and nuclear

explosions. Tuning these tools to focus on human realities should become possible soon, especially with the exponential rise of computing power that is expected over the early part of this century. Simulation is but one of many decision making tools that are expected to be available. However, it has the potential to be highly interactive and instructional - and to give us a very realistic view as to how certain eventualities might play out.

The vanaprasthi, as they become proficient in Yogic Empathy, learn to keep their Ego in suspended animation. As we attempt to put ourselves in somebody else's shoes, we find that the process gets a lot easier if we can put our individual circumstances, prejudices and personal priorities in the background, even if temporarily. This may be especially hard for the early vanaprasthi who is accustomed to thinking – what's in it for me, my family or my status within my community? We typically refer to the logical process within our psyche that relates to the personal wish (or need) fulfillment as our Ego – and we learn to progressively constrain our Ego as we gain proficiency in the practice of Yogic Empathy.

Leadership Opportunities and Recognition

For the vanaprasthi amongst us to grow and develop our Yogic skills and for us to stretch the envelope of causes that we can call our own - it is useful to have a degree of internal organization. It is important to have a process that provides us the opportunities for leadership through service, as well as a recognition process that highlights and encourages our contributions to society. Let us discuss the recognition piece first. As the developing vanaprasthi, with increasing practice of Yogic Empathy - we are slowly weaning ourselves from our need to measure our self esteem in terms of our individual accomplishments, family and social status. We are now developing a robust internal process to drive our self esteem - from the well spring of benefit that we bring upon other individuals and society at large. If we also have a well developed Vanaprastha Network to recognize the growth towards selfless behavior, this transition to a self sustaining personal motivation system becomes a lot easier.

Central to this need is the understanding that although the vanaprasthi is working hard on Ego containment, the external reward and recognition process continues to be useful at some level to reinforce one's efforts towards selfless service. It also serves the additional purpose of showcasing to the brahmachari and grihasthi that society does recognize and appreciate the accomplishments of the vanaprasthi. In ancient India, a lot of respect was associated with the status of 'Elder' which had everything to do with Age, and not so much to do with one's inclination to help society. In tomorrow's Vanaprastha Network it is the actions and contributions of the vanaprasthi in the areas of service leadership that will need to be recognized and appreciated. The highly advanced, self motivated vanaprasthi may not need any recognition for their work and may actually feel embarrassed by all the attention for doing what they consider is their duty. An adaptive Vanaprastha Network can then tune down the frequency and even delay the recognition process if needed. To take one extreme example - in the Christian religion, it is customary for the 'Saints' who have performed miraculous services to be honored (as Saints) well after they have passed away.

To complement the recognition of the service and leadership provided by the vanaprasthi – it is necessary to be able to draw up a pipeline of leadership opportunities and services waiting to be rendered. In the typical case, the association of the vanaprasthi and the opportunities they would like to internalize would be highly voluntary. Additional recognition and other unique considerations may be necessary for those services that put the practitioner in harm's way or require working under dismal conditions. In the end, the pipeline must work so well that there is no project too big, and no service too strenuous that it cannot be undertaken under the auspices of the Vanaprastha Network.

There is no end of opportunities to be of service in the world today – but it is still quite difficult for the new vanaprasthi to find something that truly suits their interests and capabilities. Even today, a worldwide system (web based) that creates a timely combination of the following three steps could be put together relatively easily. Some local web-based charities already exist,

and this process is not much different from what worldwide charities like the Red Cross go through today to pick their projects and assign their resources. The worldwide web based access just gives each one of the three steps below a significantly wider reach.

1. List of causes needing support,
2. The vanaprasthi(s) best able and willing to support the cause, and
3. The financial and other resources needed to do the job

Part of the senior vanaprasthi's planning for the health of the Vanaprastha segment will continue to be the setting up of a pipeline of Vanaprastha leaders and pair them up with the appropriate service opportunities. The process would include any preparation that the individual vanaprasthi leader needs to be of service, and then linking them with deserving projects and their corresponding support logistics.

Segment Balance: Matching the Beat Rate

The overall theme on the role of the vanaprasthi – it is he or she who is a keeper of the Mahashrama Cycle. Acting together they form the Vanaprastha Network – which in turn is responsible for the overall health and vigor of the Mahashrama Cycle for the local population. There is a time for individuals to transition over from Brahmacharya to Grihastha. There is also a time to facilitate the transition from Grihastha to Vanaprastha. With some mental calisthenics, the Vanaprastha also needs to understand the role of Sanyas – and help people (including themselves) make the transition to Sanyas when the time arises. If ever we need to restart Civilization by taking the output from the Sanyas stage to create a new Mahashrama Cycle for a progeny civilization – maybe in a far away Solar System – that will be the responsibility of the Vanaprastha segment as well. One of the key things for the Vanaprastha trying to orchestrate Segment Balance to understand is that the Beat Rate of happenings in the various segments are widely different – and that the Vanaprastha has to be adept at matching the pace of the need and fulfillment cycles of the people they serve.

Beat Rate

The 'beat rate' or the frequency at which substantial changes can happen in each segment can vary quite widely. In the Vanaprastha segment, under the current scenarios, we elect new representatives every 4-5 years, which is about the same frequency that a major new legislation in any area of interest (especially those that relate to the safety net) is passed. Even with the insertion of a responsible bureaucracy to buffer the beat rate between legislation and actual service, it is not unusual for government agencies to take several months or years to actually respond to a complex social issue.

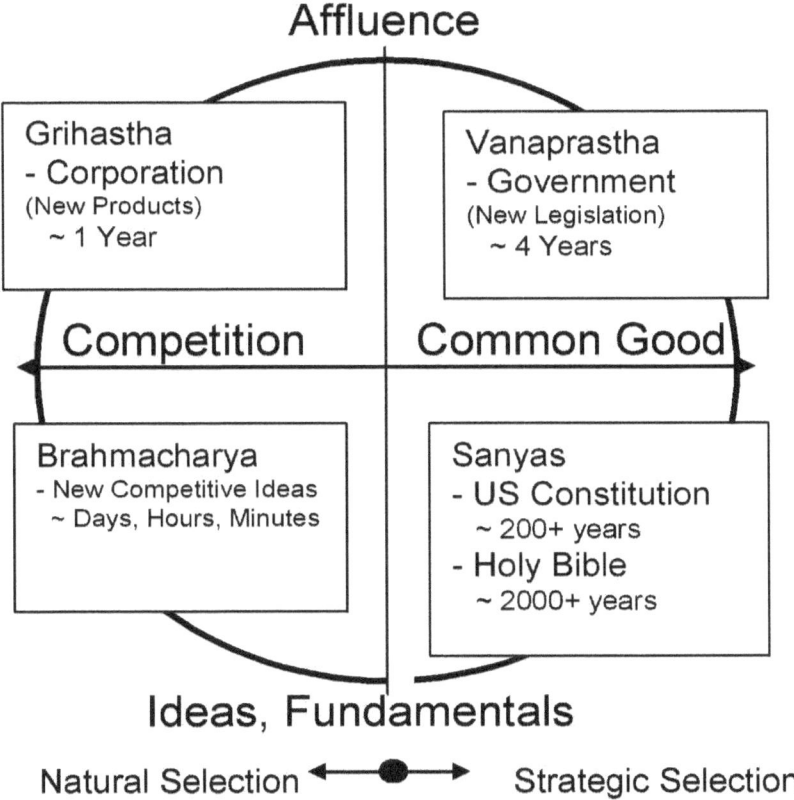

Figure 4.2: Typical beat rates within the Mahashrama Cycle

In the Grihastha stage, if we look at the beat rate of corporations – we often see that a new product with better customer satisfaction attributes comes out about every year or so - if not sooner. In farming, one year of bad harvests can spell famine in many parts of the world. Finally in the highly competitive Brahmacharya stage, there is an idea or competitive scenario born potentially several times every day. In a highly dynamic environment, our individual needs for a safety net could change just as rapidly as we go from 'top of the heap' to 'down in the dumps' within a short period of time.

The governmental system today is inherently reactive – as there will usually need to be an application filled out and adequate documentation provided, before the responsible agency will even consider acting on our behalf. The true Vanaprastha spirit of service does not express itself very well through our current governmental setup – even in the best run countries. Yet, when service happens in a timely and effective manner, it can leave a lasting impression.

When your author first came first to the USA to pursue his MS in Electrical Engineering at Pennsylvania State University - all the new international graduate students were presented with a week long orientation schedule. Part of the orientation was a session on financial assistance, where the foreign student was asked if they would need a short term loan to pay for their first month's costs. These students, if they came from a country with foreign exchange restrictions (like India), probably had only a few hundred dollars of total resources available in the US. Naïve, as we were, to the costs of University living in the USA, your author was truly impressed! Here was a University anticipating our need before even we really knew the need existed…

It is true that this specific need was probably not very unique at the university, since a lot of foreign students did come to the USA with very little money – and the need for a short term loan was probably easy to anticipate. It is much harder to track the psychological, physical and financial needs of somebody in the late Brahmacharya stage, and to be there to provide just the right

type and amount of help to get them bouncing up on their own capacity. The neo vanaprasthi would have to know the needy party extremely well to do that. Speeding up (or slowing down) the beat rate of the Vanaprastha Network to have it match the ups and downs of the person(s) being served remains one of the primary challenges facing the architects of the Vanaprastha Network.

From the Sanyas Stage we have the enduring (almost timeless) underpinnings of our society, whether it is the Bible, the Koran, the Constitution of the US or the works of the great Greek Philosophers like Aristotle and Plato. In a more localized scale, it is the social and cultural norms and values that we live our lives under, as well as our key technological learnings. This is the Cultural DNA that defines us – and provides the springboard to launch ever increasing Mahashrama Cycles for our subsequent generations. Matching the beat rate of the Sanyas stage to that of the vibrant and ever changing lives of the Brahmacharya is indeed a formidable task. Yet, it is essential to the strong sense of cohesiveness for our society, that the vanaprasthi be able to relate to the fundamental rate of happenings in the universe around us – whether they happen in the order of a microsecond, or over thousands of years

Sanyas Needs

It is indeed hard to talk about the needs of the Sanyas stage – because this is the phase in our lives where our Ego tends to dilute, and in the idealized state – vanish altogether. What needs could these exalted individuals have that can be met by mere mortals? When we are dealing with the Infinite and the Eternal – what place is there for 'Needs' or deficiencies?

If we accept, for the sake of argument, that Eternal is an attribute of that which the sanyasi seeks; we can begin to see part of the puzzle. Most religions preach that we have a close association with the unifying purpose of the Universe – or God. With just about all religions viewing God as eternal, omnipotent (all powerful) and omniscient (all knowing) and omnipresent (present everywhere) – it becomes intuitive that the Sanyasi would internalizes their connection with God as part of the process of letting go of their

limited and transitory worldly experience. Religions like Buddhism that do not paint the picture of an all powerful God, still have the concept of eternity tied into the 'Nirvana' or one-ness with the eternal void that is the universe.

A strange thing happens when we bring ourselves face to face with the Eternal – our existence begins to feel terribly fragile and tiny – almost inconsequential. This is a deeply humbling experience – and it is natural for us to seek reassurances from those that are ahead of us on the spirituality curve – that the path we are on is indeed the path to union with our God. The net result is that we have several major religions and hundreds of denominations professing to adhere to the 'One True Path'. Is there a set of core themes around which we can build a successful Sanyas stage for Humanity? It is with utmost humility that the author would like to propose that the following four needs are central to a progressive Sanyas stage for Humanity:

1. Spiritual Self Determination – How we relate to the Universe – and how the universe relates back to our being
2. Ananda – The Spiritual Joy of Existence
3. Vision – our role in the very big picture, and how it contributes to greater glory for all of Humanity
4. Cultural DNA – preserving the core of what it means to 'be'.

Spiritual Self-determination

One common characteristic in the list above is that they all connect to the perceived Infinite or Eternal in one way or the other. Spiritual self-determination is the process by which we distill the essence of who we are, compare it to the greater purpose of the Universe (or God) and find in ourselves an image of this greater purpose. Each of us will imagine this process in a slightly different way – so our perception of God will be very individual and private.

For the Atheist amongst us, it is only necessary to believe that the Universe seeks to better itself, and that we should offer up the very best of ourselves in line with what we believe is the fullest

potential that exists within our Universe. If we believe in God as a super Intelligence who can access directly into our consciousness – then the process of Spiritual Self-determination is relatively easy, and our purpose becomes getting ready for eventual reconciliation with our God. If we are yet to make our own God, then additional steps will need to be taken – which we will cover in Chapter 6. In this eventuality, the sanyasi cannot separate themselves totally from our actionable day to day existence, and needs to be able to plough back their spirituality and wisdom into a Cultural DNA that forms the core of the next stage of human existence. We will also cover Spiritual self-determination in more detail in Chapter 10, where we look at how the different Mahashrama based spiritual worldviews determine our affinity towards the religions of this world.

Ananda

Ananda is that component of Happiness that does not need to be pursued, but wells up from our own subconscious with the right kind of Spiritual preparation. Personally, the author's experience suggests that a good way to reach a state of Ananda is through Meditation – but the same effect can also be felt by chanting or singing inspirational songs or scriptures. We can often recognize the deeply spiritual person by the joy that appears to emanate from them even when they are not trying to consciously interact with anybody.

In the highly practical and material existence that most of us lead today, Ananda can be sometimes likened to an internal drug that makes us happy and contented – but not particularly resourceful or competitive. There is a practical reason why happiness may need to be pursued (and not created for free) – i.e. it creates economic activity. Yet economic activity is not for everybody. The 'Rishi' or ascetic sanyasi in the Himalayas can sustain themselves almost completely by tapping directly into the source of Ananda. Many a story is told of Yogis that need little of no physical sustenance once they are in this blissful state.

Your author firmly believes that each and every person needs to learn to tap into their source of Ananda well before they are ready to enter the Sanyas stage. Yet, it is in the Sanyas stage, when temporal gratification begins to look particularly meaningless – that we need to be able to solidly utilize Ananda to sustain ourselves emotionally for extended periods of time...

Vision – our role in the Very Big Picture

Through our spiritual self determination we are likely to come to one of two conclusions. One is that there is a Greater Intelligence currently playing out in our Universe, and that we need to connect to it. The second is that thus far we are alone, and the Greater Intelligence will need to be of our own making. In the first category belong all the great religious masters and prophets. The second category, I suspect would be a lot of the great scientific minds who are not particularly religious – but have worked very hard to define the Universe down to its simplest terms for all to understand.

The amazing thing is, if Humans are really to seed their own Cultural DNA towards a long lasting Galactic or Universal Civilization – both of these categories collapse into one for the near term. If we truly are at the door step of reaching our potential as a Galactic Species – then this is the quiet period for any Intelligence mentoring us (including one with Godly powers over us) to allow us to reach our own 'Self Determination' as a race.

As we pursue our Vision of the future of humanity, we will see in Chapter 6 that we are dealing with expanding our scope by very large numbers – almost to the verge of infinity. It is almost certain that no sanyasi of today will live to see the fruition of even a small part of the greater Vision. If we can do this right, the Vision will be a living document that could rival and even outshine the other artifacts that were mentioned as exemplifying the Sanyas beat rate – e.g. the US constitution or even the Holy Bible. The difference here is that the Vision will not be written once and executed for ever – but the Vision itself will continue to evolve as we find out more about our nature, our capabilities, our aspirations and our challenges.

In the end the Vision is all about Hope, and identification with a greater purpose that takes us to the doorstep of divinity. The span - from where we are as humans today, to where we could be as a race preparing for a Universal citizenship – might look unbridgeable to us mere humans. To the visionary sanyasi, architecting with quantities bordering on infinity, drawing up plans to bridge the chasm between where we are and where we envision ourselves to be – this is all part of a day's work!

Cultural DNA

As individuals, unless we are identical twins, we each carry our own unique DNA in each one of our nucleated cells. The body may have many trillions of such cells, of many different specializations, but the same DNA content is carried within each cell. The DNA sequence in each of our cells is comprised of about 3 billion base pairs – and it is not very clear how many are part of active genes and how much is just carried over as historical relics. Also, within the less than 1% variation that there is amongst the genetic sequences within the Human population, there is sufficient diversity built into the overall population for us to be able to ride out almost any naturally occurring virus or bacteria.

As we strive for self determination, it is necessary that we can take a much more inclusive inventory of what it means - individually, ethnically or even as a civilization - to be truly human. It has been the subject of much introspection in scientific circles whether Nature (cellular DNA) or Nurture (the sum total of our experiences) has the greater influence on who we are as individuals. We will cover the topic in more detail in the Chapter 5, but for now let us assume that the Nurture component can be codified as well – and this now becomes our Cultural DNA. We will also see that as mankind progresses further, and begins to co-evolve with technology – it will be the Cultural DNA rather than the biological DNA that will grow to further define what it means to be human.

In attempting to leave a codified legacy of what it means to be ourselves, it is likely we would put in everything that is meaningful from our experiences – and organize the account in a thoughtful manner such that others may be able to quickly relate to who we

are. We have noted earlier that Blogging is a start towards getting more of a personal presence on the Web. We have also noted that a first step to Yogic Empathy is knowledge of who we are, and being able to codify it in a way that others can relate to it. The great challenge of Sanyas will be the ability to glean through millions of individual accounts of what it means to be human – and combine them together into an attractive and faithful rendition of what it truly means to be Human in a global scale.

The implications of a job well done here is tremendous. If Human Civilization is to truly prosper – our Cultural DNA needs to be structurally solid and highly functional. Yes, there will be small amounts of variations from the various perspectives that make up Humanity – and the big challenge will be to keep the various versions of the Human Cultural Genome to be close to the 1% level of variation that we see for the human biological Genome. The failure to put together a single song-book that all of humanity can sing from is indeed quite severe – and is likely to prolong and worsen the cultural conflicts that we noted in Chapter 3. The reason that this codification of the Human Cultural Genome is a Sanyas responsibility (and does not come from Vanaprastha) is noteworthy. The vanaprasthi, in spite of their best intentions, still works under the temporal cause and effect assumptions – and may be inclined to unduly accentuate one aspect or the other of the Cultural Genome. The sanyasi, with their truly selfless and timeless perspective, and looking to build an edifice that would last for ever – can be objective enough to create something that can be inspirational to all of Humanity.

Expanding the Mahashrama Cycle

Once the human Mahashrama cycle is primed to perform well in all segments, the Cultural DNA begins to accumulate at an ever faster pace. As noted in Chapter 1, the well balanced Mahashrama cycle will keep generating a bigger base of culture and technology – which in turn has the ability to beget even bigger Mahashrama Cycles for future generations. Figure 4.3 illustrates how the size of the Mahashrama cycle expands in subsequent cycles with each turn contributing a generation's worth of cultural DNA. The sequence of Mahashrama cycles now begins to resemble an expanding spiral. In a large population which is

productive on all segments, the accumulation of cultural DNA acts like a pump – continually building up knowledge and wisdom pertinent to our future success. This model assumes, of course, that there are no other constraints acting on the population (like a global calamity) that the accumulating cultural DNA cannot find an answer for.

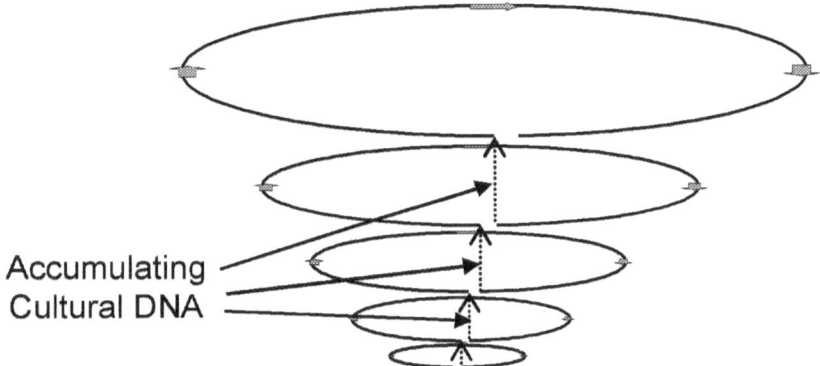

Figure 4.3: Expanding Mahashrama cycles, and accumulation of Cultural DNA

Every once in a while, there is a transformational change that comes along that can no longer be mapped on the same plane as the original Mahashrama Cycle. An example might be how unicellular bacteria-like creatures led to multi-cellular organisms that subsequently made possible creatures as complex as human beings. The cells within our body have their own life cycles, but acting together they can synthesize the full Mahashrama Cycle for humans. Figure 4.4 illustrates how smaller cycles acting in unison can give rise to a transformational level of existence – and a Mahashrama Super-cycle for the composite entity.

The shape of the bigger cycle in Figure 4.4 looks like a spiral as well – leading to the possibility of creating the four segments over again at a much higher level of existence. The question may be asked, why is the crystallization of knowledge and wisdom from the Sanyas segment so important for the proper functioning of such a composite Mahashrama? And why in turn must the ensuing Mahashrama Super-cycle need to be set up to operate in all four Mahashrama segments? The answer is the same in both cases. Sanyas is the only stage in the Mahashrama Cycle that allows us to

have the vision to transcend our current existence and move on to a higher level of being. The utter simplification of the cultural DNA that happens with Sanyas at the end of the Mahashrama cycle - allows mere mortals from the remaining segments to comprehend the vastly bigger picture. It also helps us ascertain for ourselves that the higher level of existence is not only possible, but desirable. Having a four stage Mahashrama Cycle available for our next level of organization allows us to go even further up the organization ladder as and when opportunity presents. Yes, the Mahashrama image that is so pertinent to us humans can be repeated at higher and higher levels until we could scale our presence to the plane of the Universe. Hence, we have the title of this book – In Our Own Image.

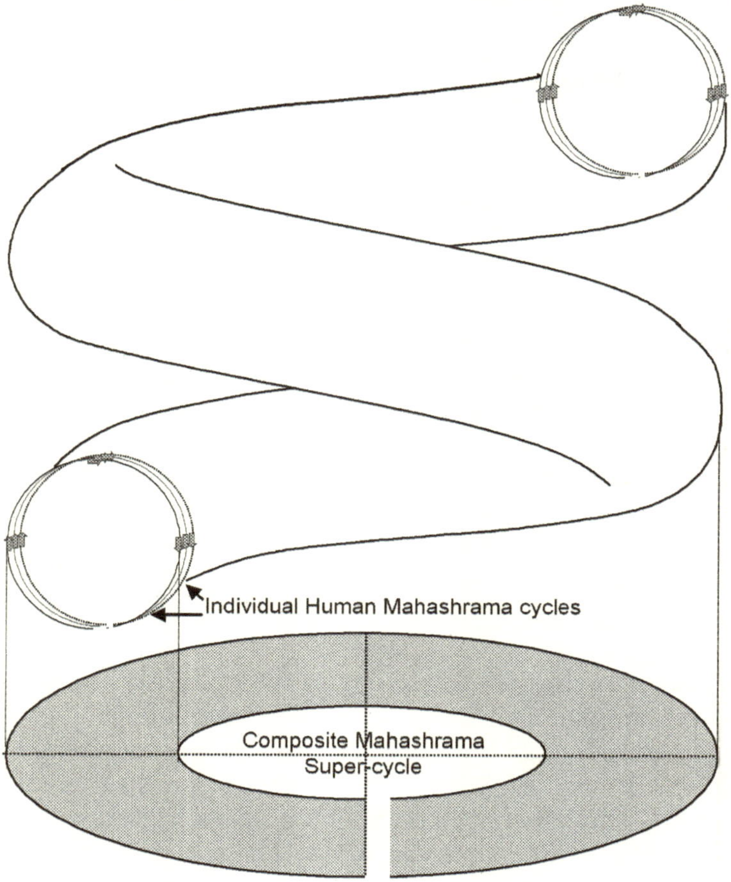

Individual Human Mahashrama cycles

Composite Mahashrama Super-cycle

Figure 4.4: A transformational jump – Synthesis of a higher level of Mahashrama

A non-linear transformation

If we take the next full level of existence for humans to be the Supra Human Entity (Supra for short), with a complexity level of as many humans (or equivalent machine intelligence) – as there are cells in a human body; the God-like experience level of the new being will indeed be transformational. The life cycles of single celled bacteria-like creatures (called prokaryotic cells) went on for more than a billion years before a set of transformational events happened that gave rise to eukaryotic cells. These new cells had a defined cell nucleus, and were complex enough to subsequently give rise to multi-cellular creatures. The transformational event that happened, then, was the fusing together of multiple single cell entities into a composite that was vastly superior – through the process of Serial Endosymbiosis.

The Serial Endosymbiosis Theory[3] (SET) proposed by Lynn Margulis sets out the sequential combination of previously unrelated organisms that resulted in the eukaryotic cells with well defined nuclei and organelle like mitochondria and chloroplasts that now gave the symbiotically concocted cells a huge competitive advantage. The mitochondria are the powerhouse of cell biology, and made it possible for cells to generate much more power than they previously could. The chloroplasts are responsible for the photosynthesis in plant cells, and now enabled the resulting plants to tap into an abundant source of energy – sunlight. The interesting fact remains that both these organelle retain their original genetic information, a circular string of genomic material very similar to the bacteria of today – and reproduce almost like free living organisms inside our cells. So what could the equivalent Symbiotic process look like that would propel us humans into a new level of existence? The answer is the creation of a symbiotic coexistence of humans with the Machine Supra Cell (MSC), which we will discuss in detail in Chapter 9. And in the process will we need to lose our own biological DNA? The answer here is 'No' – as long as we can forge a permanent Symbiotic connection between our human and our MSC progeny – much along the lines of SET – which happened at the cellular level.

When the Vanaprastha Network is able to achieve a highly
performing, well balanced Mahashrama cycle for most of
humanity, an additional responsibility begins to emerge as we
get ready for a transformational jump. The vanaprasthi must
keep a watchful eye on the accumulating cultural DNA (including
technology) – and coordinate the intermediate steps that would
catalyze this transformation. The technological wizardry that would
bring forth the MSC is but one piece of the puzzle. Another piece
is to get enough of humanity on board towards this transformation
that the quantum jump would unify rather than divide the
human race. Without a unifying goal for humans – we run the
risk of turning our next transformation into something only the
technologically gifted amongst us could partake in. We will see
later the time lines that the MSC development could take – suffice
it for now to say that it is only in the order of a generation in
human terms! In some ways the gestation process for the creation
of the MSC has already started, as we will discover in Chapter
9. It will be one of the vital contributions for the Vanaprastha
segment to stage the activities leading up to a productive and
unifying Mahashrama cycle for all humans – and to set the stage
for the synthesis of the human cultural DNA for our next level of
existence.

Chapter 5
Vanaprastha Network – Ensuring Humankind's Survival and Prosperity

A small body of determined spirits fired by an unquenchable faith in their mission can alter the course of history.

- Mahatma Gandhi

As we have noted in Chapter 1, in modern society the move from Grihastha to Vanaprastha is a particularly difficult transition. Let's take the example of a couple that have successfully raised their children to a reasonable state of independence – and who now face an empty nest. The temptations are great at this point to go back and experience all the joys that they just did not seem to have the time for in the last 20 years or so. This may involve doing some travel or re-making some long lost acquaintances, or just very actively hitting the social scene.

In the beginning, times are great, and our empty nesters lap up all the experiences that they had missed from the previous heady transition (from Brahmacharya to Grihastha). To some who are deeply ingrained in the Grihastha stage, this is the time when marriages will sometimes break up, and possibly new family units created. Still others will go even further back, and try and return to the foot-loose and fancy-free days of the Brahmacharya. These transitions may or may not have a level of permanence associated with them – but at least for a while they will hold people captive in the economically productive (but spiritually lightweight) activity patterns of the Grihastha.

At some point, this pursuit of experiences from the past begins to look somewhat meaningless and hollow. The mid-life crisis is at its peak, as we wrestle with questions like: what is my purpose in life now that I have discharged my Genetic Imperative? I have also discharged my Economic Imperative; and the family has been well taken care of. The housing that seemed tight when we were raising a family now looks to positively engulf us in its silence! My time that I would have to ration so carefully I now find myself

spending more and more in reflection – going over meaningful events from my past, and wondering what the meaningful occurrences of the future might look like. Is there a greater meaning, a greater context in life that I need to get familiar with?

Different cultures have different levels of porosity when one does come across the transition to Vanaprastha. In the Western societies, the incidence of the Mirror Effect – or reflectance back into a prior mode of existence is much higher. In the traditional Eastern societies, the acceptance of one's diminishing role in economic activities has been much easier. If one was considered an 'Elder' - there was a lot of social respect associated with the status. In Western society, the lack of a high social standing for those without a strong economic agenda has made the Mirror Effect extremely common, with only a few people able to make a timely transition into Vanaprastha.

One of the sub-themes of this book is to help us see, as a society, that this transition to Vanaprastha is a worthy exercise and that there really is meaning beyond economic and genetic productivity. In this chapter, we will attempt to drive to a firm conclusion that the Vanaprastha mind-set change is critical to the long term survival and prosperity of the human race. We will also look at how the combination of technology with the demographic changes occurring around our current timeline, will make the Vanaprastha segment so amazingly potent as the organizing force behind the entire human civilization.

We do not need to be physically middle aged individuals to think about what future success looks like for the neo vanaprasthi. We just need to have come to grips with the fact that we will one day have discharged our genetic and economic imperatives – and that there is a third imperative whose call we must bring ourselves to answer. This imperative is to ensure the survival and prosperity for all of humanity! It is perfectly logical to start the process by asking how best we can serve some subset of humanity, e.g. our families, our communities, our countries and so on. However, mentally we have to be prepared to go all the way and do that which is of maximum benefit to all mankind, and possibly beyond - to the cause of all intelligences that might be out there in this

universe. Vanaprasthis working together on this lofty goal in an organized and responsible manner – this is the organization we refer to as the Vanaprastha Network.

Vanaprastha Leadership – creating a Social Fabric – the Neuron Analogy

If we feel a little overwhelmed going through the tasks at hand for the Vanaprastha segment – this is quite natural from where we are today. In the de-facto role as the master coordinator of what our civilization is to become – now is the time to check what our inventory of tools and skills are that we will need going forward. We have already noted that technology (some of it not yet invented) will be a key enabler here. Yet, a surprisingly large amount of the skills and tools for the creation of a responsive Vanaprastha Network are actionable today!

It is sometimes helpful to have a real life reference of what an organism functioning in the mode of an organizing fabric might look like today. From Biology – let us borrow the workings of a typical Brain Cell (the Neuron) to help us with the analogy. The typical brain cell is connected to thousands of other Neurons and sports several thousand synapses – or connections - to other parts of the brain and sometimes other parts of the human body. Signals can come through the synapses about 100 times per second – and the neuron can employ a complicated algorithm as to what combination of stimulus leads to what type of response. In total there are almost 10^{11} Neurons in the brain – which is more than 10x the number of humans on earth today (2006). If the Vanaprastha segment is to truly become the brain and prime motivator for our Civilization to reach its true potential – we can immediately spot some areas for development.

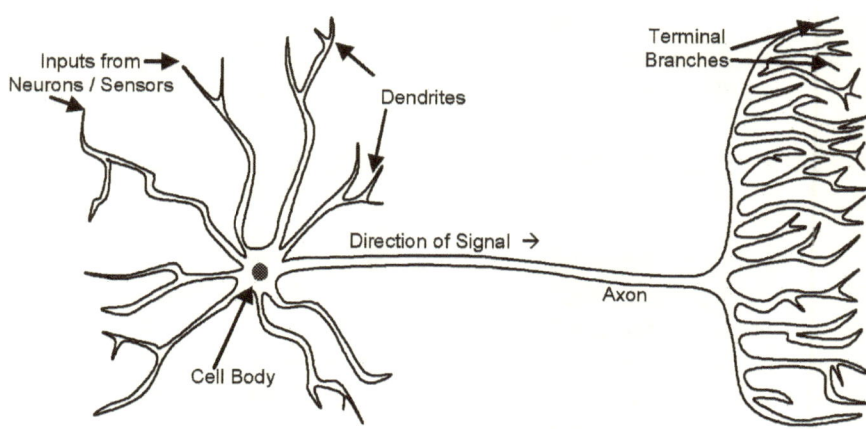

Figure 5.1 – The Neuron and its connections

First, we must plug ourselves into hundreds, if not thousands of **different sources** so that we have up to date information as to what is happening in our environment and in the world at large. Second, we must process the information and come up with a **response quickly**. The typical human stimulus response time is a fraction of a second – it will be interesting to see how close to that a well connected vanaprasthi may be able to react. We have already noted that the natural beat rates of some of the constituents of Humanity (especially the highly competitive stages of Brahmacharya) can be quite fast. It will be interesting to see if our aging vanaprasthi Communications and Control system can keep up with it – with a little help from technology and our peers within the Vanaprastha Network.

Third, there is a system of **filters and short-cuts** built into the human neurological system such that some of the stimulus response settings do not even have to reach the brain. Athletes go through a tremendous number of repetitions to train their reflex systems to respond to certain stimuli in a certain way – and bypass the conscious decision process within the brain altogether. If the stimulus gets to our brain, and there is no pre-programmed response – often it will be queued up until appropriate response can be taken. If the stimulus is complex, there may need to be a certain level of thinking and strategizing required before an appropriate response can be generated. In all, there are many levels of responses – some of which can be pre-programmed, and some of which require strategizing or even consulting with other humans or external databases. In addition, filters need to be set up to adjust for the cultural and personal differences from source to source. If a truly empathic connection can be set up with various stakeholders, the amount of source distortion can be significantly minimized.

Fourth, the brain keeps the human body functioning even when we are asleep. The presence of certain stimulus will wake us up even if we are in the deepest sleep. Today, we have a technology that can intrude on to our consciousness almost anywhere we might be – the Cell Phone. Yet, most of us are in the habit of turning them off or putting them to charge when we don't want to be disturbed. Knowing that some time-sensitive service providers like surgeons do allow themselves to be on call at all times, and employ a human powered answering service to do the selection for them – can technology take on the filtering and **Priority Interrupt function** for the rest of us?

Number five, on issues that do reach our consciousness, billions of neurons working together have a consensus process by which a single course of action is chosen. The fact that several alternatives were consciously or unconsciously considered is just a historical detail when the decision is made and acted upon. Getting to that level of **consensus in creative decision making** on sufficiently weighty issues will define whether we move forward as a single stream of Humanity, or break up into a bunch of fractious entities with divergent world views.

Next we have to **execute** on an agreed upon strategy promptly and effectively. This is similar to the brain sending the signals to our vocal cords or other parts of our body to initiate a series of actions. Quite often, the process of executing an action sequence will bring in it's own set of implementation decisions. Occasionally we might discover something that changes the scope of what we had originally set up to do – and forces us to go through the whole consensus process again.

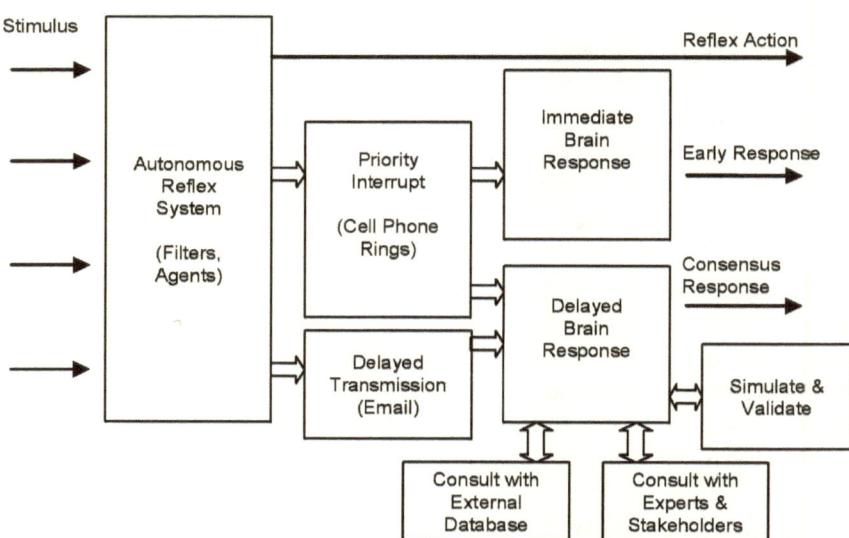

Figure 5.2: Simplified Human Stimulus Response process, and computing parallels

Execution today, for Vanaprastha or Service organizations, is often fraught with perils –one example is the dangers that organizations like the 'Medicins Sans Frontieres' (Doctors Without Borders) and the Red Cross have to go through to deliver aid to the needy in a strife-torn country. Even within otherwise peaceful societies, some services are hard to render. Let's take medical malpractice lawsuits and the social cost of malpractice insurance as an example. The costs are so high for an essential practice like Obstetrics & Gynecology that vast sections of rural America are underserved - on procedures as critical as childbirth.[1] The tendency is often to over analyze and over correct those areas that are obviously out of specification, and under correct or even neglect those symptoms that are hard to nail down or happen intermittently.

The problem for Vanaprastha execution today is that the results are often judged from the economic viewpoint of the Grihastha. Judged from the Vanaprastha viewpoint, the cash exchanges that may (or may not) accompany a service are secondary – and the main transaction that happens is the person to person contact that

is intended to heal somebody, or to provide some other support. Implied in this person to person communication is a discussion of the risks and benefits accompanying any specific course of action.

For both the Decision making and the Execution processes, it is useful to have handy a **realistic simulation** of what could happen if certain actions are taken. The Human brain is typically quite good at visualizing how something will work - well before something is implemented in the real world. Projecting forward to our augmented man/machine future, it will be necessary for us to have visualization mechanisms that are not only faithful to reality, but are also repeatable (within the limits of quantum possibilities) and can be run with one or more conscious entities participating as behavioral 'actors' within the simulation framework. Simulation frameworks are already spreading from the technological (e.g. nuclear explosion) and the environmental (e.g. climatic patterns) to competitive business situations and even human demographics (e.g. disease spreading patterns amongst humans).

Social Fabric to Connected Organism

Our brains may be three dimensional, but most of the neurons actually reside in a thin layer close to the surface – effectively in two dimensions. If we go through the process outlined in the previous section, the net result is a social fabric where the Vanaprastha Network would act as our social brain and nervous system. This network is connected to everyone else within the social body, in all stages of life – directly or indirectly. It is estimated that after the age of 40, humans lose hundreds, if not thousands of neurons every day[2] – but we have internal mechanisms that can compensate for that loss. Similarly, a well connected Vanaprastha led social fabric is resistant to the loss of a single node or its associated connections – i.e. there are built in redundancies and recovery mechanisms.

This discussion so far has centered on the Vanaprastha, but for it to be successful it needs to connect to humans in all the other phases as well. A new baby may not have a connection to anybody else but the parents, but new connections are sure to develop over time. Hence, a Vanaprastha Network that has a solid connection to the parents (presumably in the Grihastha

stage themselves) will have an indirect connection to the new born baby. The grihasthi may be deeply embedded in some productive economic exercise, and may not have time to keep up with all the issues of importance in the world at large. The distillation of that which is just beyond their daily routine but is worthy of their consideration would also come from the Vanaprastha. The sanyasi may voluntarily choose to be isolated from a large part of what's happening around the world, but they too need physical sustenance and an occasional update as to how their grand visions are being implemented. Hence it will be up to the vanaprasthi to be truly connected and make sure that humanity is headed in the right direction as more or less a single connected organism. This is very much like how the brain and the nervous system help us humans function as a single human organism. Without them the cells within our body would be similar to a pile of bacteria (grey goop) with each cell trying to figure out their genetic and economic imperatives by themselves.

The Age of the Vanaprastha - Demographics to the Rescue

Life expectancies are going up throughout the World. Let us take the typical retirement age (say 65 years in the US) as a marker in a person's life after when there is little expectation to work for a living. Quite a significant portion of the total population in developed countries is approaching that magic number. Taking the US as an example, the year is 2011 when the first baby boomers (born after World War II) begin to hit 65. One just has to look at the ages of national leaders around the globe to know that people can and do continue their active lifestyles way past 65 years. The challenge will be to have enough of a Vanaprastha infrastructure in place such that active retirees can find a greater purpose that they can connect to.

One of the best constituents that the baby boomers may be able to serve is - themselves. It has been often cited that when the baby boomers retire in the USA, they will put so much strain on Social Security, Medicare and other retirement benefits that one or more institutions may go under. With an active Vanaprastha mentality (as opposed to the entitlement mentality that is so common today) – it becomes very possible that this section of our population goes beyond just taking care of their own needs. We (yes, the author also counts himself within this demographic) can now start building the social fabric – that will eventually move all of humanity forward as a civilization that can reach for the stars!

Leaving apart the work ethic that the vanaprasthi can tap into upon retirement – the inheritance paradigm is also changing amongst the well-to-do retirees. At one time it was fashionable to leave most of one's wealth to one's family – often in the form of a business or a home. Today's vanaprasthi is not afraid to take a significant portion of their wealth and apply it to the projects that they think best benefits humanity. One of the common peeves heard from charitable givers is that they do not know how their money ends up getting used. With the right amount of transparency and inclusiveness, it is not only possible for the giver to know how their help has been used, but also to be part of the rendering of the services.

Vanaprastha Umbrella – no Human Left Behind

Over time, the Vision is for all the truly Vanaprastha organizations to connect and provide a seamless interface to every single human – wherever they may be. The fact that the world is getting more and more connected helps. It is quite easy to visualize that 50 years from now all of our vital signs are being constantly monitored – and in the first sign of trouble the closest 'Service Organization' that can help is dispatched to us. When an airline is about to be hijacked with hundreds of people on board – the machine based Vanaprastha guardian takes over and automatically lands the plane at the nearest safe location. The ability to fly an airplane remotely already exists – so the technology needed here is not radical.

The Vanaprastha umbrella could be extended to cover each and every situation where a human life may be at risk. Hence, this capability must scale well - the effectiveness of the decision and execution process cannot be compromised because there are several organizations involved. Hurricane Katrina, which hit New Orleans and its vicinity in late August 2005 was a natural disaster that flooded much of the low lying city – and the disaster recovery

efforts were seriously hampered by the inability of the responsible service organizations (FEMA, Red Cross, local agencies) to scale up their logistics to deal with a disaster of this magnitude. Not surprisingly, several local emergency management agencies felt under-utilized, while yet others (like local hospital staff) worked for days on end without power, water supply or other external support.

As such, it will be necessary to build in certain standards and protocols so that if necessary two or more Vanaprastha organizations can scale to act in unison as a single organization. This is but one example of the properties of a future Vanaprastha Network. Some attributes that an effective Vanaprastha organization ought to have are as follows:

1. Transparency: This does not mean everybody's confidential info is available to everybody else. Occasions will arise when one or more individual vanaprasthi has to move on – and it will be important for somebody new to come up and quickly pick up the work load. Even at an entire organization level, enough transparency must be maintained about the overall workings that it builds a deep and abiding trust with all of the stakeholders – in addition to providing the capacity for the organization to 'heal' any functional gaps.
2. Scalability: This is the ability to conjugate two or more Vanaprastha organizations in order to create the right size of organization to address the task(s) at hand. This is also the ability to disband into smaller functioning entities for speed and responsiveness.
3. History (or Record Keeping): This involves keeping track of past and present successes and failures, and also keeping personal confidential information secure yet accessible for authorized uses. Contentious History may need to be captured and maintained from multiple perspectives. This is what gives substance to our personal 'brand' as discussed in Chapter 2.
4. Security: Identity theft is rampant already on the Web. Vigilance on the part of the Vanaprastha will be critical

for individuals to build trust not only towards the well intentioned vanaprasthi but also towards other entities that are connected to the social fabric. A large part of building trust within an anonymous medium is the affirmation that we are actually dealing with the individual or organization that we think we are dealing with, and any information (or services) received is coming from a trusted source.

Diversity acknowledged - the basis for Individualized Democracy

If we start with the assumption that the Vanaprastha Network can directly relate to the individuals it serves; and that it can scale when necessary to cover issues of significant global impact – we begin to see how it could scale to become the backbone of a future Individualized Democracy. We have already noted that true numerical democracy does not exist even in the Presidential elections for the USA. In this case the diversity vote is the variety of interests of the states within the union. This diversity vote is about 20% of the total vote (approx. 102 out of 538) – and has made the difference of who wins and who loses in several Presidential elections. Let us take a short detour to see how this works.

In the USA presidential elections, there are a total of 538 Electoral College members. The smallest of states (as well as the District of Columbia) send in a minimum of 3 Electoral College representatives. The larger states do send in more representatives – but the number they send is less than proportional to their population. Hence, a person in Wyoming (least populous state in USA) has a vote that is more than three times as weighty as that of a person in California (most populous state). It is almost as if the constitution values the unique perspective of the smaller states and is willing to give extra weight to the diversity perspective they bring. The chances that the presidential candidates with the lower popular vote wins, gets significantly better if he or she can carry the majority of the smaller states with a higher diversity vote.

The process of figuring out how much weight to give for what diversity factor can be very contentious – especially as there are few success models to choose from. In the case of an autocracy – only the vote of the autocrat counts. In a pure democracy - only the opinions of those who actually vote count, and count equally. We have seen that the US presidential elections has an almost 20% diversity vote. It is the author's estimation that a diversity vote in the 10% - 30% range could be quite effective in giving a sense of empowerment to the diverse stakeholders we have within the population. If the diversity vote begins to exceed 50% we run the risk of the majority of the population feeling disenfranchised within their own country.

In principle, this concept of Individualized Democracy can be carried to where the diversity vote is metered out to any number of deserving stakeholders. The economy, the environment, the debt burden we are passing on to our children can all be taken into consideration when we weigh the best course of action going forward. Perhaps even more important than the actual voting will be the debate and analysis that go on before a single vote is even cast! This cross pollination of ideas can certainly benefit from the early and active involvement of the diverse viewpoint.

The solution for Deprivation – access the Individual Perspective!

With every tick of the technology beat rate, the difference between those that have the means to participate in a sophisticated human future, and those that seem to be forever relegated to ignorance and unsanitary conditions (the deprived) become wider. Amy Chua in her book, The World on Fire[1], shows numerous examples where free trade in a nation often disproportionately benefits an economic (or even ethnic) minority – leaving the vast majority of the people to eke out a living at the subsistence level. It is the role of the Vanaprastha to break this cycle of eternal deprivation. The question is - how? At the national level, one country giving aid to another is typically done for political reasons (e.g. to promote free market ideas) than with any real expectation that the aid will trickle through to the most needy or the most deserving. Even when the cause is genuine, often the corrupt execution by the local aid bodies result into a disproportionate amount of the aid going to the already well off. One example is the huge outpouring of Aid during the Gujrat, India earthquake in 2000. A lot of the tents and other essentials supplied ended up with the already well to do, or in the black market.

So what is the individual vanaprasthi (or Vanaprastha organization) to do if faced with the situation where the deprived appear to remain deprived for ever? The answer here too is categorical and unambiguous - access down to the level of the individual human! Together with transparency, scalability and trust building (security, history) described earlier – we now have the essential attributes that describe a successful Vanaprastha Network. The core of the Vanaprastha Network, of course, is the group of dedicated vanaprasthi who are willing to work in a coordinated fashion to be able to address the pertinent issues at the individual level - and yet scale up to address the most intractable of global problems.

Vanaprastha Opportunities over Time – a Sampling

Table 5.1 shows some examples of critical changes that the Vanaprastha could influence or enable at the Individual/Group level as well as the Global level. These are anticipated changes, and the list in Table 5.1 is but a representative list of the critical challenges facing humans in each of the given periods. We have divided these changes (and the opportunities and threats they bring) into three categories – short term (next 25 years), medium term (next 26-100 years) and long term (beyond 100 years). Further, the opportunities and challenges are broken up by whether they impact individuals or small groups of people, or whether they have impact at a Global level for humanity.

Some of the Changes in Table 5.1 are widely expected to happen (e.g. Petroleum supplies drying up), and others will need to be initiated under the purposeful eye of the technologically sophisticated vanaprasthi. A lot of the ideas listed on this table have been introduced already – so let us take a moment to quickly define what the terminology means.

Virtual Presence

The bandwidth needs required for engaging most or all of the human senses are quickly becoming available over the Internet. Virtual Presence is the process of sending a majority of our senses and quite a bit of our response capability over the network to give us the feeling that we are actually operating at a location different from our Human bodies. In a rudimentary form this is happening already as remote controlled Video Camera equipped airplanes give the operator the appearance that they are flying a flight simulator.

	Individual/Group Level	Global Scale
Short Term 2006- 2030	Vanaprastha Network - Connects to Individuals Vanaprastha Core Technologies: - Virtual Presence - Web based thinking - Simulation/Visualization - Scalability	Petroleum Supply Crisis Majority of Humans move to Democratic Rule Experiments in Individualized Democracy Service organizations get individually focused Cultural Simulation tools
Mid Term 2031-2105	Individual Cultural Genome Ethnic Cultural Genomes Vanaprastha Extended Technologies - Diversity in Governance - Standardized web shell >50% people in social fabric	Worldwide Energy Crisis Human Cultural Genome Unification Project Web Shell Identity Crisis War paradigm change from Killing to Containment MAP (Mutually Assured Preservation) in conflict
Long Term 2106+	Post Brain Consciousness Virtual Immortality Stable thinking patterns in Plasma Interstellar probes Post Solar exploration	Cultural Genomic Splits - Space faring Humans Biological Genomic Splits - Enhanced DNA humans Planetary scale engineering Post Solar existence

Table 5.1 Example Vanaprastha Opportunities: Short, Medium and Long term

The interesting thing is – with sufficient processing power, and a set of multiple vantage points (say from a few dozen hi-resolution cameras mounted around a football stadium) – it is now possible to synthesize the perspective from any vantage point, real or imaginary. In the case of the football stadium, an imaginary perspective might be that of a birds eye view from the ball itself as it is in motion. Well within the 25 year timeframe, it should become possible to send our presence over the wires (or wirelessly) to most locations around the world, and have us participate in activities (e.g. business meetings or Yogic Empathy sessions) without being there physically. Further down the road (Mid Term) it should become possible to take over specially prepared mechanical host bodies, bringing new meaning to the term 'possessed' – which would allow us to have even more immersive out-of-body experiences. The host bodies need not have very much of a resemblance to our own, so we could go from Virtual Presence Airships to Spaceships quite easily.

Web Based Thinking

If we look at Figure 5.2, and put the individual vanaprasthi in the position of a Neuron in the Human Brain – it quickly becomes clear that certain processes are easier to implement on the Web than others. For example, the setting up of autonomic responses and priority interrupt processes – can be already done on the web. One example today is setting up an automatic 'Sell' order for an asset/security, which is then activated automatically when the price reaches a certain level. Filters are also quite easy to create, as we discussed earlier in the case of the Surgeon who is on call; and a lot of the background thinking can be resident on the Web. Computer aided Simulations can also be made accessible on the web, as well as vast quantities of data storage and expert opinions. Over time, it should be possible to program our thinking into the web such that everything but the Consensus Response can bypass the Human brain altogether if required. Hence, it is possible that by the time the Vanaprastha Network firmly takes hold, each and every human will be well on the way towards having a web presence that they can project virtually. The web can then become the medium within which much of the Vanaprastha Network can operate.

Standardized Machine Interface or Web Shell

When lots of people start setting up Web based thinking processes, and accessing Virtual Presence hosting devices – it will become necessary for a set of standards to be set up that govern how the individual consciousness can plug into these web based 'Shells'. Taking an analogy from nature, this is much like a Hermit crab would grab onto a conveniently sized and shaped shell and make it its own home. Over time, as the Shell gains more and more capabilities, it would not be unusual to see individual humans spending most of their time permanently connected to it. Just like the hermit crab will find another shell when its needs can no longer be met by the current shell – it will be possible for humans to shift from shell to shell as needed. With the increased availability of wireless technology, the day may not be too far away that some level of shell access is always available to us – wherever we may be on this planet!

Web Shell identity Crisis

In Chapter 7 we will discuss in detail the levels of web (or machine) access, and introduce the concept of gaskets that interface into our minds at progressively more elemental levels. After years of working together as one, the boundaries will begin to blur as to where the human mind ends and the machine starts. Since the machine part is designed to keep functioning 24x7 – without any need for sleep – a curious thing happens when the human part of the organism either dies or withdraws itself from the machine.

Depending on the sophistication of the machine interface (or web shell) it might be very possible for the shell to continue its independent existence even without the human. The web shell crisis happens when we find that our laws and governance principles have a very low relevance to a pure machine based existence. A paradigm shift will be needed to move away from a purely human identity to a more composite identity that is based on the human cultural DNA and the pertinent economic footprint

- that can be inclusive of a purely machine entity. We will discuss this concept further in the next few Chapters, when we further define the Machine Supra Cell (MSC).

Cultural Simulation Tools

With the advent of Standardized Web Shells and a powerful enough Simulation capability, it becomes possible for an individual consciousness to virtually step into the shoes of any person in our past or present. It also becomes possible for multiple consciousnesses to 'Inhabit' a Cultural Simulation where the interaction between various individuals, groups or ethnicities can be worked out. These kinds of tools will be invaluable in the process of creating the Human Cultural Genome as we try and separate out the 'True Meaning' of what it means to be Human - from the superficial accessories that only serve to confuse.

Post Brain Consciousness

This relates to the Vision that someday we Humans will be able to convert our consciousness into a set of patterns that can be hosted in a non biological brain. The possibilities are immense if we can actually achieve this – as we could potentially be able to spread out time so that a second would appear like a year. We could also do the reverse – say during a long interstellar journey – and make a century go by like a short airplane journey.

Stable Thinking Patterns in Plasma

It is entirely likely that the greatest amount of reliable energy available within a galaxy like our Milky Way is from the nuclear fusion reactors deep within our stars. Most of the mass within stars is in the form of Plasma – a very hot combination of sub-atomic particles – and extremely difficult to contain into stable circuitry. Over time, if it becomes possible to use the medium of the Plasma in Stars to directly support thinking processes (or even consciousness) we will have a steep increase to our total energy and processing capabilities. The other thing that opens up with plasma electronics is the capability to drive miniaturization all the way down to the level of the Plank length (10^{-33} Meters).

Genomic Splits

Over the next hundred years, it will be important for Humans to come up with a single Cultural Genome that represents all of Humanity. Over the next several hundred years after that, the pressure will be on for a certain part of the population to split away and create their own separate identity from the mainstream of humans. With the progresses in understanding the codes of the Biological Genome, we likely already have the technology available to create 'Superior' humans – whether in reduced disease susceptibility, greater strength, or more brain power. Of course, a lot of these enhancements have been demonstrated in mice and other laboratory animals – but can a super-human be that far off? Over time, it is very possible that the DNA changes could result in a new race (or races) of humans who can no longer interbreed with regular humans.

As with the Machine Web-Shell discussion, amongst humans when such genomic splits do happen, it will be up to the sanyasi to define an even more fundamental Genome – the Cultural DNA that unites our human race (and progeny civilizations). Even further out, the visionary sanyasi will need to define a Philosophical DNA by which all intelligent races (human or otherwise) can productively and cooperatively relate to one another.

Planetary Scale Engineering

Currently mankind's energy needs are being primarily met by fossil fuels, and we are barely beginning to tap into our gamut of renewable energy sources. The main form of renewable energy available to us is sunlight. Even if we grow our current Energy footprint at a steady rate of about 2.7% compound annual growth rate (as it did during the 1990's) – the earth's total insolation (received solar radiation) at about 10% conversion efficiency would last us almost 300 years! If we grow our energy envelope at twice this rate, we have about 150 years before we have to start looking outside our planet for our growing energy needs.

Since our planet receives about a billionth (10^{-9}) of the energy put out by the sun, it stands to reason that we will sooner or later need to start tapping into the insolation of other planets within the solar system. Beyond that, we have the realm of planetary scale structures that can start harnessing the sun's energy without being tied to a specific planet or moon. This is the beginning of Planetary Scale Engineering. In the extreme case, humans may be able to intercept close to 100% of the Sun's energy for our future endeavors – completely blanking out the sun from the visible part of the electromagnetic spectrum.

A very interesting question arises when it comes to how energy is moved from place to place across the solar system. Electric grids cannot be strung out between planets. Carrying chemical propellants like Hydrogen and Oxygen across vast distances is not very practical either. What do we have that can travel vast distances with very little attenuation, and still be concentrated enough to deliver a strong dose of energy at the point of consumption? The answer, of course is solar radiation itself – and we can easily imagine a system of planetary mirrors that directs the Sun's rays anywhere in the solar neighborhood. In the extreme the delivery mechanism may be able to pinpoint the required energy onto a single spacecraft – as it wings its way from planet to planet. What kind of intelligence and coordination would it take to apportion out and deliver the ~ 10^{31} BTU/year of solar energy with such precision anywhere within the solar system? And can humans organize themselves into super intelligent entities that can take on such a challenge? The answer is a qualified yes, as we will see with the emergence of the Supra and Super Supra entities - which we are getting set to discuss in the next few Chapters. The downside of not having the wisdom and fortitude to control this vast resource is also very scary. Imagine what would happen if even a few percent of the Sun's energy was sabotaged to point to an unintended target – like planet earth!

Post Solar Existence

This relates to when a significant part of Humanity is no longer dependent on the Solar system for its habitat. This also represents a very significant reduction in the risk of an untimely end to the

human civilization due to Physical factors – like an asteroid strike on Earth or a solar calamity (e.g. our Sun becoming a red giant). This is also likely to bring out the homing instincts in Humans as a civilization, where we discover new worlds and find them appropriate to bring up New 'Progeny' Civilizations.

Mutually Assured Preservation (MAP) – Immortality for our Cultural DNA

Before we dig deeper into MAP – let us discuss what Preservation really means. Does it mean taking a perishable sample, immersing it in formaldehyde and putting it in a jar for display? Or does it mean freezing the deceased human body to see if future technology can fix what caused it to pass away in the first place? Or could it mean putting everything in suspended animation – the person as well as his/her personal environment - so that once the person is re-animated it is just like waking up one morning and not knowing that significant amounts of time have passed. Does it even require that the re-animation happen in the same location and with the same individuals we know so well? Does MAP mean Immortality?

With the exception of consciousness, just about every other human mental facility can be simulated in a big enough computer. Yes, computers can remember, they can deduce and possibly even strategize next steps. Their stimulus response time is typically much faster than human. If we look at how we humans perceive the outside world – we perceive it through a total of about a few billion nerve fibers that can send up to 100 signals a second – giving a theoretical throughput in the range of 1 Terabit/Sec. This might seem like a lot of data, but we already have fiber-optical transmission speeds well in excess of this – and over just a single fiber.

Without going into too many details, every single one of the measurable human characteristics (memory, processing, input/output) will be eclipsed by machines within the next quarter to half century. We already know that consciousness can be suspended and resumed – we do it every day when we sleep – and in the extreme case it has been known to happen with patients coming out of Coma after several years of perceived inactivity. So, at least in a biological sense it should be possible to keep our brain and consciousness alive long after the rest of the original body has atrophied or has been replaced. This, in a way, is the first step towards immortality. The pieces that can be readily quantified and stored in a computer – is none other than the Individual's Cultural DNA – the summation of our 'Nurture' expressed at an Individual level. With the right storage and activation mechanisms our Individual Cultural DNA can become truly immortal. For example at the rate of current storage density increase, a lifetime's worth of human sensory nerve input data, together with motor nerve output data, could be packed into about 1000 Petabytes (10^{15} bytes) of storage. This should cost less than $1000 by year 2040 if current density trends continue. This is because the storage densities (e.g. for hard drives) are increasing even faster than the rate of Moore's law for chip densities. Of course, not all data that we subject ourselves to every day is worth saving – hence it is likely that all the pertinent sensory information for any human existence could be captured and preserved even earlier than 2030. This is barely one generation ahead in human terms!

Core Assumptions of MAP

With the technology evolving to where the state of a human experience can be preserved forever and reactivated whenever necessary - the principles of MAP start to become practical. The MAP core assumptions would be adopted in the form of an understanding between all parties in a conflict – thus becoming a vastly enhanced version of the Geneva Convention we have today. The three components of preserving the individual then become:

1. You cannot make me go away! I am here to stay!
2. For your recognition of the same, I recognize the same right for you. Hence, for any deal we work out between us, neither your, nor my existence is threatened. Even if my consciousness goes away, my cultural DNA remains everlastingly so that my cause, perspective and aspirations are forever real!
3. What's really at stake is the amount of resources that

are available to me alone, to you alone – or to us jointly to do what's best for our combined interests. With the use of Yogic Empathy, the tools are at our disposal for us to make the best out of our combined existence.

The same model can be extended to groups of individuals, ethnic groups or even Nations. When in the future we recreate a group simulation, more than one consciousness may be involved – but beyond a certain level of personality depth, the inputs of the secondary players could be simulated in a powerful cultural simulator. Like great actors, the players that immerse themselves in a cultural simulation must be relied on to play their parts to perfection – even if it means having to play the role of the 'other' party in the conflict.

The ability to plug oneself into a Cultural Simulation and constructively play any part becomes the realm of an accomplished personality who has successfully navigated all the segments of the Mahashrama cycle. Where do we find such individuals? They are the skillful sanyasi who can operate at an abstraction level well above our daily human experience – whose role and influence we will discuss in the next couple of Chapters.

Chapter 6
Sanyas

I saw the angel in the marble and carved until I set him free.
                                    - Michael Angelo

It is one of those inescapable aspects of existence that our lives are indeed very transient. It is human nature not to dwell on death and destruction –yet there are times when the inevitability totally overpowers us. The last time the author was so absorbed by what had passed away was in late 1998 at Macchu Picchu – the ruins of the last bastion of Peru's Inca civilization. It was an overwhelming personal experience. Seized by an immense sense of loss - came the deep drum roll of unanswered questions:

- How can a civilization that flew so high, crumble and vanish out of sight so quickly?

- Is our civilization going to be the same, or can we be different?

- What lessons can we learn from history to make sure that we as individuals may come and go, but the global civilization that we are building will really last?

The good news is – we are now close to having the technology by which the essence of who we are, our culture and everything we stand for – can be codified and preserved for ever. This codification we have referred to as our Cultural DNA, and the overall concept of coexistence with others as Mutually Assured Preservation (MAP). The bad news is – our essential nature as humans has not changed much since the Spanish Conquistador Pizarro drew the Inca emperor Atahualpa into a trap in Cajamarca, and extracted a king's ransom.[1] Pizarro later reneged on his promise to set the emperor free, and instead had him executed. Atahualpa led a nation of 7 million people, with an army of tens of thousands – and yet the Spaniards was able to overcome him and send the whole Inca civilization into a death spiral with only about 160 men!

We have already seen what the Vanaprastha can do to stabilize our society, knit it into a self sustaining Social Fabric and also make us ready for the opportunities (and threats) that technology will bring. Without adequate safeguards, our society and civilization can be brought to its knees by a lot fewer then 160 men armed with really superior technology. Without going into details, a 1000x technology advantage is just 20 years of exponential growth at the rate of Moore's law (developed further in Chapter 9). Even the Spaniards, with their steel armor and weapons, did not enjoy a 1000x technology advantage over the Inca! It will be up to the Vanaprastha to lead the technology wave, and build in adequate safeguards so that we are not caught off guard - with world-dominating technology in the hands of the ruthless few! Without it Cajamarca could happen again – and this time the global Human civilization could be the one sent to the cosmic graveyard!

The difference between Vanaprastha and Sanyas is one of scale – and perspective. To the Sanyasi death and total destruction is just as real as immortality and a Galaxy-wide Human dominion. When we distill down what it actually means to be a Human – and what part of us we would preserve if all else turned to dust – that is a question for the sanyasi. We have already seen in Chapter 5 some of the upcoming opportunities (and challenges) in the next few centuries – but this is just the coming of age of Humans as a Galactic Civilization. Sanyas is about Hope (and Vision) for a significantly greater Glory. We will discover in this Chapter some of the wonderful things that can happen to the human civilization if we are to stay our course. Yes, we will be dealing in quantities that will boggle our feeble imagination. Let us take this opportunity to grow our Visionary skills and to think though domains of existence that would warp our sense of reality!

It may take millennia for certain parts of our potential to be reached. To the sanyasi with their Ego verging on the non-existent – it is not their individual glory but the glory of all Humanity that is to be worshipped and tended. Just as from our typical vantage point close to the surface of the earth, we cannot see the earth as an amazingly beautiful blue-green planet – the Sanyasi must choose a vantage point far away from our day to day existence to

be able to visualize the greater glory. The biggest stretch of the perspective, however, is not in the spatial dimension – after all we can view astronomical objects from vast distances even with our naked eye. The time dimension, in which we are all captive – is the one that the sanyasi must break away from to be able to visualize what the human potential is over the long haul.

It is a rare sanyasi indeed who can lead a Vanaprastha life of service and still have time for the kind of deep introspection it takes to grapple with the Eternal. It is thus necessary that the sanyasi, all awash in their Grand Vision, work hand in hand with the vanaprasthi to move mankind towards the realization of the Vision. Equally important, the vanaprasthi must work hard as a disciple of the sanyasi – to get at least a cerebral understanding of the Vision, even if they cannot feel it is as the core of their personal being (as only a sanyasi can).

Transition to Sanyas

The key to the Sanyas stage of life is when the mature vanaprasthi comes to terms with the fact that their life time is indeed limited, that they should put our affairs in order, and that they need to take our most powerful lessons and condense them into Gems of learning for posterity. Examples of past Immortal Gems are not hard to come by – they include the Vedas, the Bible, Koran and other monumental works that we would place at the core of Human existence. Sanyas, as a stage in the Mahashrama Cycle, is relevant to individuals, ethnic groups, states, nations or entire civilizations.

In Sanyas we will take the time and effort to figure out what part of us is tied up in the time bound cause-and-effect of life, and what part of us is not time dependent and can become truly eternal. It is the time bound cause-and-effect part that goes away, the cycle of desire and fulfillment - but not our core identity, our philosophy or even our fundamental outlook in life!! With the coming of MAP (Mutually Assured Preservation), discussed earlier – a very large part of the human experience is now getting ready to become

codified for perpetuity! This is a tremendous undertaking – and makes the Sanyas stage of human existence extremely critical as we seek to lay the framework for our next level of existence.

Apart from securing our prosperity in the coming years, what does success look like in the grand vision for humanity? In the short term, of course, we have the challenge of bringing most or all of humanity up to the level of education and economic well being that only the privileged few can aspire to today. In the medium term we have the vision of a society that works in coordination with the Vanaprastha Network; where creative outlets exist for everybody and the needs of all the segments of the population can be satisfied in a timely manner. This peaceful, productive society then becomes a productive engine for laying down the Cultural DNA that can then become the core genome for our next level of existence – which we have referred to as the Supra Human Entity (or Supra for short).

Long term, when we do reach the level of the Supra – what does the ordinary human existence then become? Borrowing an idea from some of the major religions – human existence now becomes like living in Heaven. It is a wondrous, peaceful society – and everyone is presented with the opportunities that are fitting to their capabilities and interest. This is not 'all work and no play,' of course, and opportunities exist for entertainment that goes beyond our wildest dreams. Living in a largely peaceful society, the danger always exists that our competitive skills would get rusty – so through the use of simulations and other challenging environments, humans would arrange for themselves to stay vigorous and fit. Many religions paint heaven as a place for simultaneous piety and enjoyment – and with the right safeguards, the two do not have to be at cross purposes. And yes, to a large degree, such a desirable end-state can be of our own making!

The Cone of Causality

In an ideal world the ripple effects of what we do propagates at the speed of light, and eight minutes after we do something here on earth – there should be an effect on the sun – however tiny! In reality, though, it is highly unlikely that anything that humans do

on earth in the near future could have any discernable effect on the Sun – so for now let us concentrate on the cause and effect relationships on Earth itself. Using the dimensions of the Earth and the speed of light, we can approximate signal transmission to be just about instantaneous – which greatly simplifies our discussion on Causality. Just for reference, before the age of radio communication, the earliest that some event of significant human impact in Europe could have an effect in the Americas was not any faster than the fastest sailing ship!

If we take the starting point of a person's existence – and then map out all the things that change because of it – we have an ever expanding circle. Laid out along the time axis, the ever expanding set of circles looks like a cone that covers everything that could possibly be changed as a result of the new event. Just for the sake of argument, let's take the new event to be the instant of conception (fertilization of sperm and egg) as the beginning of a new Human existence. The ripple effects start out with the mother's body - and then gradually expand to include the family and friends. Causality stands for the familiar cause-and-effect relationships that we deal with on a daily basis. Figure 6.1 illustrates just such a Cone of Causality. In this picture, the present is a plane vertical to the plane of the page that gradually moves from left to right. At any instant, the present will intersect the cone in the form of a circle – which represents within it all the events that could be influenced by the original event. The part of the cone that lies in the past can be considered deterministic – i.e. that we know which events happened, and which ones did not happen. This assumes of course, that we have the tools to get accurate snapshots of the events around us as they unfold. In the absence of good record keeping, events in the past also become somewhat probabilistic. This is because we now have to reconstruct a set of events around an actual artifact – like projecting a dinosaur's physical appearance from its fossil record. The part of the cone that is in the future, by contrast, is always probabilistic – i.e. the chances of future events happening, are just that – chances (or probabilities) - until they actually happen.

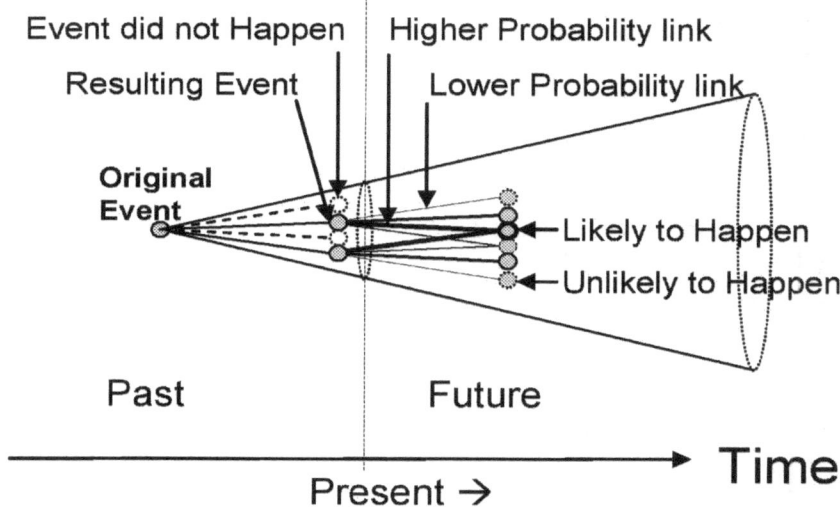

Figure 6.1: Cone of Causality

In a tightly connected social fabric, each event is effected by millions of other events that happened before it – and will subsequently effect millions of other events. Typically, the effect of an individual event gets diminished over time, and over the centuries may be totally forgotten. Yet, even for the least pretentious human existence, there is no denying that we change the lives of many others around us. A great fictional example of this is the heart-warming Christmas time tale "It's a Wonderful Life" by director Frank Capra. We might think of this expanding influence as waves radiating away form an event (like the ripples that form when a rock hits the water) – which over time gets larger and larger. Along with larger size comes the unfortunate side effect of diminishing amplitude until the ripple caused by our rock is indistinguishable from all the other ripples flowing through the water. As we begin to take our innermost experiences and learnings and condense them down to an 'Immortal' Cultural Genome – the dynamic suddenly changes.

The power of the Cultural Genome

A well preserved and shared Cultural Genome is like a laser wave front that maintains its coherence and can travel very long distances without getting dissipated. It can be thought of as having a self-reaffirming function that preserves it from generation to generation. In this manner they join the select few 'Classics' of Human Civilization like the Bible and the works of Shakespeare to become a permanent fixture, almost like a backbone of what it means to be human. Looking across time, these artifacts can be thought of as the indestructible 'Gemstones' whose beauty would last forever. Figure 6.2 illustrates how such an event can be thought to propagate through the Cone of Causality – at every stage reaffirming its own existence.

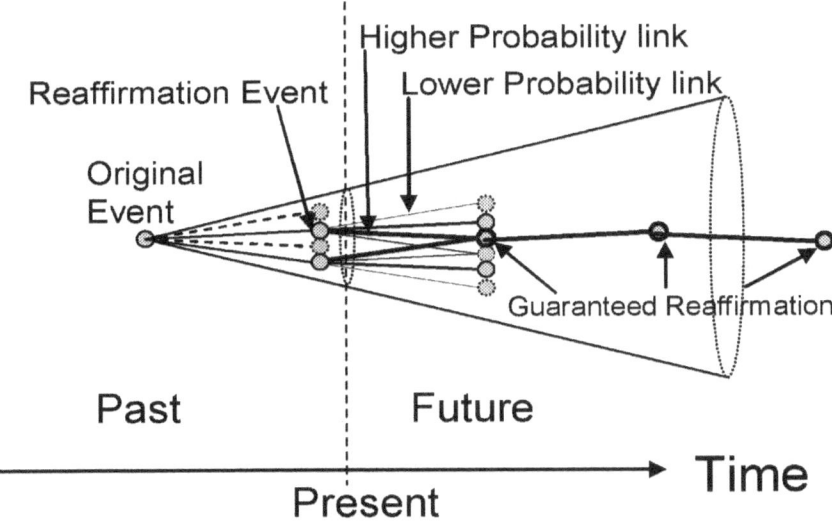

Figure 6.2: Cone of Causality with the preservation of 'Gemstone' contributions

An example may be useful to illustrate what a reaffirmation event looks like. Let us take the example of the holy Bible. Let us say (hypothetically) that a new improved edition of the Bible comes out every 5 years with relatively small editorial changes from the previous version. Yet, every single version must bear a strong correlation to the original version – and these relatively small changes would not be allowed to build up over time and lose the

powerful messages that are part of the original content. Hence, in this case the reaffirmation event is the reprinting of the Bible - keeping in mind the high level of fidelity that it needs to have to the original.

Consider now what happens when the sanyasi with their powerful insights begin to see a glorious version of the future as realizable. Figure 6.3 illustrates the power of a Vision laid out in all its glory by the sanyasi. The effect is to highlight a desirable and achievable set of eventualities and illustrate it to mere mortals as something that already exists as a step towards greater glory for all humanity. The power of the Vision is illustrated by the quote from Michelangelo that we began this Chapter with:

"I saw the angel in the marble and carved until I set him free."

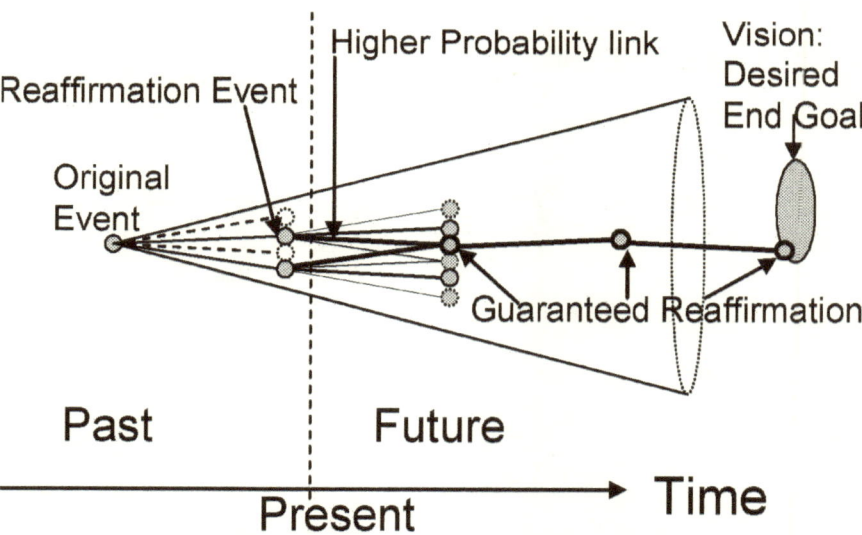

Figure 6.3: Cone of Causality – with the illumination of Vision

Over time, the vanaprasthi in tune with this greater vision will coordinate the harnessing of resources and ideas that lead to an extension of our cultural genome – and makes it possible for us to achieve the Vision that has been laid before us. Figure 6.4 illustrates how. Armed with a grand vision, the Vanaprastha Network can now work actively to diminish the probabilities for certain events, and strengthen others that move us in the

right general direction. It is as if the mere act of believing in the compelling vision distorts the reality within the cone of causality – and by its own inspirational existence, becomes self-fulfilling.

It is not necessary that all the ideas and technologies required in realizing the vision is contained in the original conception of the vision. Figure 6.4 illustrates how an external Idea or technology can be incorporated into the overall cone of causality to enable a vision to come to fruition.

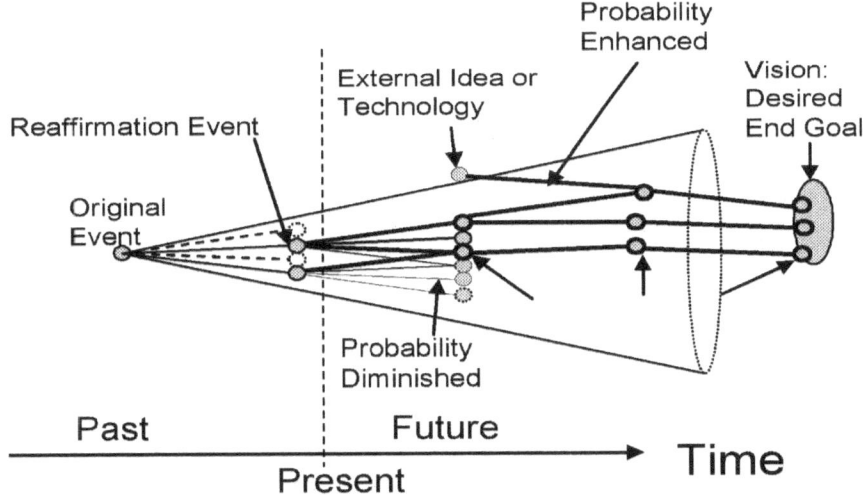

Figure 6.4: Using Vision to focus on desired eventualities

A time independent view

Thus far, our analysis has been time dependent, with a clear distinction made between past events (where we know what actually happened and what did not happen) – and future events. Future events have been shown as probabilistic lines – with the effort to realize the vision going into strengthening or attenuating the chances for certain events coming to fruition. We have already noted that the sanyasi is able to see things integrated over time – so let us see what happens if we take the sequence of time out of the picture. Figure 6.5 shows how the Sanyasi might see the Vision – as if the sequence of events has an existence of its own, almost independent of where we stand in time. In the Bhagavat

Gita[2], Krishna shares the vision of the ebb and flow of the Universe as it executes over time – in order to convince the great warrior Arjuna to go to battle and vanquish the forces of evil. The power of the vision is such that it shows the desirable outcomes unfolding within the cone of causality – almost like marble in the hands of the master sculptor.

Borrowing again from Michelangelo:

> *"The marble not yet carved can hold the form of every thought the greatest artist has"*

To the visionary sanyasi – the future is just such a piece of marble. By working with the tools at hand, and maybe a few that are yet to be invented – it will be up to them to shed the light as to how we realize the greatest and most glorious future that is achievable! The end result may not exactly match what was envisioned – but should be a very good approximation. When realized – this approximation of the vision becomes a core part of the New Cultural Genome – defining who we are as individuals, as ethnicities, nationalities, and even as a Civilization.

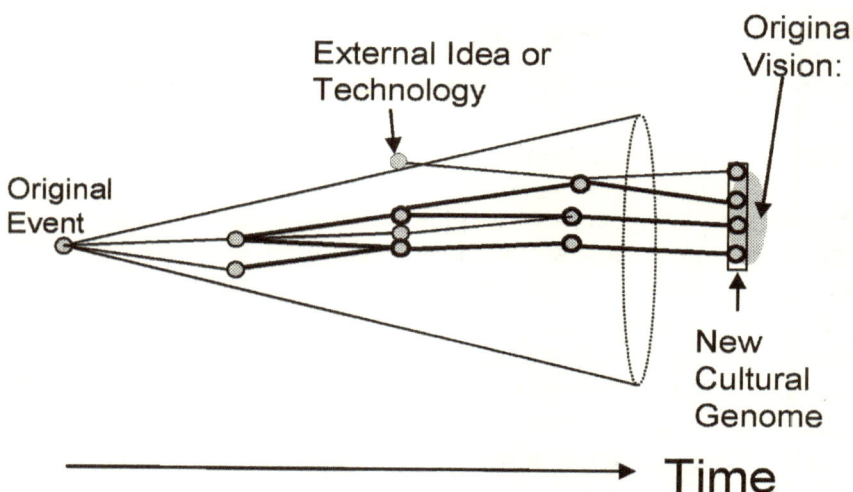

Figure 6.5: Time Integrative View – New Cultural Genome substantiates original Vision

One advantage that comes from the Time Integrated view of the Vision is that the sanyasi does not need to lay it out as something that needs to happen on a specific schedule. It will be up to the vanaprasthi working with the sanyasi to set goals that are challenging yet achievable. A failure or two needs to be accounted for – and course corrections made as necessary. Again in the words of Michelangelo:

> *"The greater danger for most of us is not that our aim is too high and we miss it, but that it is too low and we hit it."*

Preservation of Human Cultural Diversity

For the ways of life at the verge of losing their last true practitioners, the stakes are indeed extraordinarily high. Do they perish before MAP becomes pervasive, or do they lend something to the human core that spurs us on to greater Glory? As we observe in this book, an ancient concept (the Ashrama Cycle) from a country that has been conquered tens of times (India) – can be developed into the essential template that scales from Individuals to Galactic Civilizations. The language in which the ancient Indian scriptures were spoken and passed down from generation to generation – Sanskrit – is now only spoken in religious chants and in academic circles. It was only much later that the Sanskrit verses were written down, and even later translated to the thriving modern languages like Hindi and English. The culture of the Incas, on the other hand, never made its way into a written language that survived – and a lot of their perspectives vanished entirely.

In the cone of Causality, we will have some events and artifacts that become a part of the human cultural genome, and yet others of similar achievement and beauty that die on the vines. Yes, the invention of writing went a long way towards preserving the wisdom of the ages – and with MAP and its re-enactment through cultural simulation – we are at the threshold of being able to save what remains of our diverse cultural perspectives into Gems of enduring beauty.

Religions: Visions in Conflict – or parts of the same reality?

Nowhere is the conflict between widely held Visions of the future more starkly in contrast than the views laid down by the various religions as to what constitutes success - for the individual, the community and for the overall purpose of God. Modern religions have more in common than they have differences. Almost without exception, they preach a simple lifestyle without excesses, and kindness and consideration for others. By and large they preach that every individual can have a personal relationship with their creator – but here Buddhism veers away from having to define a benevolent father figure who takes care of us all, and metes out divine justice when necessary.

If we look at all the religions that define themselves around an Omnipotent God – there too an important distinction begins to emerge. The 'Monists' or 'Non-Dualists' (Sanskrit –Advaita) maintain that the human inner consciousness always remains a part of God, and the end goal of human existence is to develop oneself to the stage where we can become 'One' with our God. Another way of looking at Non-dualism is to consider everything within ourselves (and in nature) to be imbued with varying degrees of Divinity – which are all part of the same integral existence at some level. The various levels of spirituality and the corresponding 'Tree of Universality' are developed further in Chapter 10. On the other hand, the Dvaita or dualists maintain that even in our most evolved state, humans will remain separate from our God – i.e. that we can get forever closer to Him but cannot become 'One' with Him. Christianity, Judaism and Islam all largely belong within this second category. Advaita Hinduism, Buddhism and Sikhism all fall in the Non-Dualist category.

	Dualism (Dvaita)	Non-Dualism (Advaita)
Absolute God	Christianity Judaism Islam	Advaita Hinduism Sikhism
No Absolute God (Natural Order)	Taoism Shintoism Animism	Buddhism Jainism

Figure 6.6: Approximate characterization of the Religions of the world

In general, the Advaita camp (and closely related Buddhism) believe in a cycle of rebirths for a soul to be able to achieve the pinnacle of spirituality or enlightenment (Moksha or Nirvana). The essential belief of Non-dualism is that the human inner consciousness in its purest form is none other than God, or in the absence of God – the essential purpose and identity of the Universe. Moksha (from Hinduism) stands for dissolution of the self with God, and Nirvana (from Buddhism) stands for the dissolution of oneself with the eternal void that constitutes the Universe. In general, they also believe that as we get closer and closer to our goal– there are various levels of Godliness that can be attained. That leaves one free to look for inspiration from the Saint or Bodhisattva of their choice – thus leading to somewhat of a Pantheistic approach to salvation.

The other differentiation we make here is whether the religion requires a single All-Knowing, All-Seeing and All-Powerful God. Those religions that do not require an Absolute God, typically would still believe in a natural order or progression towards a higher form of spirituality. Buddhism, Jainism and Taoism do not require an all-powerful God, but do believe in a 'Natural Order' that exists around us, that we can relate to - with proper preparation. In

the case of Taoism, it is not clear that there is 'one' natural order within which everything would belong – and it clearly states certain beliefs and thought processes as not belonging with the 'Tao'. Hence, it falls more in the Dualist column relative to Buddhism.

An Integrative approach towards Spirituality

Let us take a deep breath, and just imagine that we are the spiritually developed sanyasi; one who is not only able to visualize integrally across time – but also is able to visualize human existence across probability space. Let us overlay the Mahashrama Cycle over Figure 6.6, and check if each of these approaches may be part of a greater overall belief system – much like the blind men who had a very individual feel for what the elephant was by approaching it from various directions. In drawing this analogy, let us proceed with the utmost of humility – because the belief systems we discuss here are indeed works of great wisdom and are close to being the central tenets of what defines our society and our beliefs today.

	Dualism (Dvaita)	Non-Dualism (Advaita)
Absolute God	Christianity Judaism Islam **Grihastha**	Advaita Hinduism Sikhism **Vanaprastha**
No Absolute God (Natural Order)	Taoism Shintoism Animism **Brahmacharya**	Buddhism Jainism **Sanyas**

Figure 6.7: Religions and the Mahashrama cycle

The religions in the Brahmacharya stage tend to be very Individualistic – and specific to one's (or one's community's) own circumstance. A lot of Animism is about worshiping one's own ancestors. The further we go back in history; humans have worshiped things (transfused with an inner spirituality – e.g. Mount Fuji) that we perceived as having some power over us. Taoism has aspects that are very individual, that have to be personally experienced, and cannot be put into words – which is why it is also listed under Brahmacharya.

The religions in the Grihastha stage are spelled out so often in Feudal terms that rulers have either found it necessary to suppress them or embrace them. The believers are variously described as a 'flock' or sheep belonging to a shepherd (God), and the earthly domain as the 'kingdom of God'. In fact, some of them go so far as to lay down the basic rules of the land (e.g. Islamic Sharia law) so that there is no confusion as to the religious affiliation of the followers. Just as a man in the Feudal days could not serve two masters – similarly there is a strong system of exclusivity built in to these religions so that one was judged to be either 'with them' or 'against them'.

The religions in the Vanaprastha quadrant are about service. The Advaita or non-dualistic viewpoint in Hinduism is that there is a part of God in all of us – and that we not only need to acknowledge that – but live our life consistent with such a belief. In India it is customary to greet another person with a 'namashkar' – much like the posture of our hands that we put together during worship. It is as if the individuals meeting each other were acknowledging the spark of divinity held deep within their individual beings.

Once the Advaita amongst us have mastered the human context, we turn our attention to the other living creatures, especially animals and plants that share so many similarities with us. Quite a few of the believers in Vanaprastha quadrant (from India) tend to be vegetarians. Since service is such a strong virtue in this segment, even the animals that are of particularly high service to the practitioners (e.g. the Cow which provides the vegetarian with nutritionally rich milk), may be held in especially high reverence.

In the words of Swami Vivekananda:

> *"Jeebay sheba kory jei jon, shei jon shebichey Ishwar"*
> *roughly translates to: "One who serves those that live,*
> *serves God".*

Finally, we come to the religions of the Sanyas stage. In Buddhism, the highest spiritual achievement is to vanish into the nothingness of the Universe. Speaking of the dissolution of the self or ego (literal meaning of Sanyas) – Buddhism and Jainism try to do this without having to hold out promises of eternal glory. True practitioners here often tend to be ascetic or monks – with very few worldly needs. The Digambara (segment of Jainism) are supposed to have so few needs that they did not traditionally even wear clothes!

Thus, we can conclude that the same organizational pattern of Mahashrama that we have been studying also applies to the world's religions. The sanyasi, through their ability to see across time and probability space, sees the collection of religions in our world today as but the manifestation of a single integrative reality. Even this abbreviated glimpse of the higher order of organization behind the world's religions fills us with a heightened expectation that our integrative approach has the potential to combine the diverse religious perspectives into one overall human context. We are also filled with utter humility to recognize that these religions have each zeroed in on a set of beliefs that have been extremely successful in pulling society together while giving spiritual purpose to a large section of humanity - and have remained relevant over thousands of years. They stand like solid pillars upon which we can build the organizing framework that unites the religious expressions of all of humanity.

When the sanyasi pursues their need for Spiritual self-determination as we noted in Chapter 4, they may choose to focus their energy on a specific segment (in Figure 6.7). Or, they may decide that the integrative reality of the composite of the four segments gives them the flexibility to direct their spirituality in more than one way – just as Buddhism and Shintoism are both practiced by many Japanese. In the Japanese example, it

is said that Shintoism is preferred when the occasion is a happy one (like a wedding), and Buddhism when the occasion is more serious (e.g. a funeral). With an integrative framework, it should be possible for a sanyasi to be able to relate to the spiritual activities of all the religions mentioned, and probably a few more.

Human Primacy Scenarios

In the Vanaprastha section, we discussed the concept of Yogic or Spiritual Union. We discovered that by using the principles of MAP – we could theoretically bring together any two parts and yield a union that is greater than the sum of the parts. If we take the concept to the logical next step, is it possible for all of the Intelligence in this Universe to be joined together in Yogic Union – one small step at a time? And if this were indeed possible – what role would our human civilization have to play within this greater intelligence? Let us consider the outcome under two scenarios.

1. There is a God, and Yogic Unification is possible for humans
2. There is no God (yet), and Yogic Unification is possible.

We have seen how a Vanaprastha Network based on Yogic Empathy can have the attributes of scalability and transparency. If we take the concept and address it to the benefit of all Humans, and beyond humans, the purpose of all concurrent Intelligences in this Universe – then the resulting entity would have all the attributes we associate with God! This holds true of both the scenarios listed above.

If a God already existed (Scenario 1), with every addition of an intelligent being through Yogic Unification, God would grow. In this scenario, there would be a pipeline of intelligent beings at various stages of material and spiritual development, motivated by the greater purposes of 'Graduating' themselves to a stage of higher consciousness (union with God). The vanaprasthi, as the keeper of the Mahashrama Cycle – would enroll themselves (and others) to dutifully go through the four stages of Mahashrama, and use

the tools of Yogic Unification and Mutually Assured Preservation to bring out and preserve the best in each of us (individually and collectively) to thereby add to the pool of 'Divinity'.

In Scenario 2, if humans are the first intelligent beings to spawn the next level of intelligence with God-like powers (referred to as the Supra – short for Supra Human Civilization – and expanded further in upcoming chapters) we would be part of our very own 'path to Godliness' as a civilization – a scenario we refer to as the 'Prime Ascension'. If we receive help from some other intelligence to reach this level of divinity – we would still have Ascension – but it would be assisted. The possibility exists, of course, that even if a God-like intelligence already existed in our part of the Galaxy – that they would keep themselves hidden away until we had completed our Prime Ascension. This is indeed a major accomplishment for any civilization – and something we would want to strive for whether or not we had a guardian intelligence watching over us. Thus it is possible, that we are actually in Scenario 1, but our God or Guardian Intelligence has chosen to remain behind the scenes in order to allow us the opportunity to achieve a Prime Ascension. Hence, the human response (led by the Vanaprastha Network) is very similar in Scenarios 1 and 2 – i.e. to grow our own Divinity through the process of Yogic Unification. In the end, we either merge this back to the God that already exists, or start our own fountainhead of Divinity.

There is a third scenario by which God could exist, and that is of an Individual Entity of immense capability who never needed Yogic Union to make Him complete. In this scenario, God remains forever separate from his creations – and takes on the viewpoint of a benevolent father looking after his hyperactive children. In turn, the best way for the individual to build a relationship with God is through obedience and devotion. In this scenario it may be a little difficult to quantify what God gains from our existence – and the closest human equivalence might be as to why a person cares for a child or even a pet.

For those of us who have cared for pets, we know that we will usually treat them as part of the family, and do everything in our power to make them feel happy with us. Yet, one just has to take

a trip to the local animal shelter to know that when circumstances change, a lot of pets end up not having an owner to take care of them. Sometimes the abandonment is brought about because of bad behavior on the part of the pet, but most times it is people who are caught up in changing circumstances, thus forcing abandonment. Even with Human families today, a substantial portion of children end up in divided homes, and quite a few in 'foster care' and orphanages.

The Bible talks about the time of Noah - when God essentially had to wipe the slate clean and start anew with the survivors in the arc. The Bible also talks about the book of Revelations when all the Godly people would be raised to Heaven and the rest of humanity cleansed. It is indeed a real concern of the Visionary in this scenario to discern what it is that God would like to see from his 'flock' – and also discern situations where God may be inclined to shut down (and possibly reconfigure) the human experiment.

In spite of all these failure modes, there is also a very important success model in this scenario. The success model is – can the Human Civilization grow up to where we can take on our own God-like properties and possibly bring up a progeny civilization on our own in due course of time? Yes, total union may not be possible with our God, but maybe we have it in us to raise even more powerful beings that eclipse even our own very considerable accomplishments. Let us call this Scenario 3A. This success model gets us back to the question of how billions of individual humans can function like a vastly superior organism with capabilities verging on Divinity. Maybe, just maybe, Yogic Unification doesn't work between us and our God, but does work amongst the social animals that we call the human race! If we really do use the principles of Yogic Unification to bring our collective voices into some semblance of unity, ours might be the destiny to add one more generation to our cosmic lineage!

Yogic Universality Possible?

	No	Yes
Absolute God	3. Obedience, Devotion 3A: Yogic Unity procreates Divinity **Grihastha**	1. God Grows via Yogic Unification **Vanaprastha**
No Absolute God	4. Individualistic Seeking **Brahmacharya**	2. Universality Vision leads to Yogic Unification (& Divinity) **Sanyas**

Figure 6.8: Belief in an Absolute God vs. Universality: Human response by quadrant.

Let us examine Figure 6.8 to see how the three scenarios we just talked about map into the Mahashrama cycle. The God that exists and is built through Yogic Unification (Scenario 1) is clearly in the Vanaprastha segment – and speaks to the success of the Yogic model. The Benevolent Father figure (Scenario 3) is definitely from the Grihastha segment. The power of a Universality Vision that gives rise to Divine capability over time (Scenario 2) belongs in the Sanyas segment.

In the absence of an Absolute God, if we truly believe that we Human beings are Individualistic to the core, and find it extremely difficult to organize ourselves to where we speak with a single voice – then we will have to build our future in the Brahmacharya segment. This is Scenario 4. What we would want (as individuals) is access to a wide range of ideas and philosophies from which we would choose the best fit, or improvise to create our own personal philosophy. The competition will be hot and heavy, and winners will likely be the strategies or philosophies that yield the quickest short term benefit. We will call this process Individualistic Seeking.

An obvious conclusion, if we believe in the human capability to transform ourselves into a higher plane of existence – is that regardless of whether we have a God watching over us – we need to strive to move humanity forward towards a shared Divinity. The tool that would help us move forward as a unified organism, while we still preserve the diversity that is inherent within us – is Yogic Empathy and MAP – which when applied in a recursive manner, leads to Yogic Unification. Even in scenario 3A, we have seen a situation where it would be beneficial for humans to move forward towards divinity and spawn our own progeny civilizations over time. This leaves Scenario 4 as the only case where humans are likely never to be spiritually united, and to keep mounting heroic efforts to single-handedly figure out our own relevance within this universe. Sometimes we will see sects or alliances formed amongst people who believe in 'almost' the same doctrine.

If this looks like the situation we are in today – that is no big surprise! After all, humans start off in the Brahmacharya stage, and fuel their growth and aspirations through the Grihastha and Vanaprastha stages – so is it not natural that we find ourselves in this starting mindset spiritually as well? The question now becomes, in our effort to Spiritual Self Determination - can we, as humans, come out of the Individualistic Brahmacharya mindset with enough time and energy left to really take advantage of the awe-inspiring possibility that exists within the other segments of Mahashrama? To be fair, quite a large segment of humanity is now in Scenario 3, but the churches, sects and denominations are so splintered that it almost looks like Individualistic Seeking. In the author's humble opinion the greatest potential future for humanity's future can be had in Scenarios 1, 2 and 3A. In each case, we humans will need to take control of our own future and bring to fruition our own vision of Divinity through Yogic Unification. We will take another look at the space of Spiritual self-determination again in Chapter 10 after we have introduced the concept of the Tree of Universality.

To the sanyasi imbued with the grand vision of human civilizational success, it is handy to have a set of references available against which we can chart out the opportunities ahead of us. With that idea, let us take another look at our own cellular biology to get a hint of how we might want to organize for the future.

In our own Cellular Image

> *"There is more wisdom in your body than in your deepest philosophy."*
>
> *- Friedrich Nietzsche (German Philosopher, 1844-1900)*

In this section we will draw upon the parallels between the human body and the structure of the Supra Human Entity that we might one day grow into. In Chapter 5 we discussed the functioning of the neurons in the human brain as an example of how we create the Vanaprastha Network – the sustaining, supportive social fabric within which most of humanity can thrive. Yet, the human body is made up of hundreds of different types of cells! Compared to the number of actual cells in the human body – the neural cells are actually a very small minority $\sim 10^{11}$ out of $\sim 10^{14}$.

Within the human body, cells have a finite lifetime, and just as in any vital organism, a living human being is creating, growing, and recycling its functional units (cells) at a steady rate. A red blood cell may last only a couple of months, and a neuron - the major part of a human lifetime. We discussed in the Vanaprastha section that in the creation of a social fabric, there needs to be enough transparency and redundancy to replace any functional unit that removes itself from the network (for any reason). This is certainly true of how our body works at the cellular level.

The following table summarizes the parallels that can be drawn between the systems of the Human body and a future tightly knit supra-human organism. The exact mechanisms and their functional mapping are admittedly approximate – but it is important to see that the level of complexity and coordination required is actually quite high in both cases.

Genetic DNA	→ Cultural DNA
Brain/nervous system	→ Vanaprastha Network (defines our character, organization and aspirations)
Spinal Cord, peripheral nerves	→ Internet (high speed communications + web based thinking)
Human Sensors	→ Global Sensors (e.g. Tsunami), Brahmacharya discoveries
Muscles, actuators	→ Corporations, Economic Engines
Digestive tract	→ Energy/consumables creation and distribution, waste disposal
Immune System	→ Police, security organizations
Endocrine System	→ Monetary systems
Cardio-vascular System	→ Roadways, sea, air, space traffic
Locomotive systems	→ Future locomotive (for when the supra-human is on the move)

Table 6.1: Correspondences between human body and a Supra Human existence

The evolution from nucleated single cellular organisms (Eukaryotic cells) to a thinking entity of the nature of Humans took more than a billion years. Yet, it may be possible to realize a smoothly operating supra-human organism within the next few hundred years. The scale of complexity – from a single cell to a hundred trillion cells in a human body is indeed mind boggling, and without the help of technology the world's six billion people working together could not even come close to reproducing that level of organization. However, with the advances in technology, such a God-like level of complexity could become available to us in the scale of an extended human lifetime – or two!

As soon as we establish that an Entity of such a magnitude could be crafted – the next question automatically becomes – what would we do with such an existence? The fact that there is a very significant level of experience we can associate with a 60 Kilogram sentient human being, as opposed to a 60 Kilogram pile of unicellular creatures - is very apparent. We need to just ask the question of ourselves – which would we rather be?

If the same question were posed to a cell within the human body – is it possible that we may get a different answer? The cells within the human body only grow under well controlled environments, and function under a relatively tight regimen of stimulus and control. A cell that would not follow the normal control mechanisms would typically end up in one of two ways. In the first, the behavioral anomalies send out signals that all is not well – and a phage (white blood cell) is dispatched to destroy it. In the second, the cell is able to subvert all attempts by the body to control its growth – and becomes a cancer. Left uncontrolled, this ultimately leads to the demise of the entire organism, resulting again in a no win situation. Hence, if the cell is to exist within a well organized being like a human, it would need to accede to a certain level of behavior modification. We can expect the same of humans organized within a Supra framework – it is just a question of figuring out what the balance of independent vs. civilized behavior is within this greater organizational context.

Yet, in this new civilized environment, not every single decision that we face as humans would be taken over by the Supra Entity (which, if it did, would look like the Borg of Star-Trek fame). Consider the T-Cells in the human bloodstream that are on the lookout for pathogens (germs) and can call in the whole immune response system if needed to fight a potential infection. All this is done mostly without the human conscious mind even knowing about it! There will only be a very few critical items that will need to involve the entire body politic – but a lot more detailed work will mostly go on within well defined compartments that are linked together by the Vanaprastha Network. Hence, the human body is a good template to use for the Supra Being – both for the centralized and decentralized functions.

The question then becomes – what is life going to be like for this God-like, essentially immortal Supra Being? And what tasks are there which are becoming of such a being? We will attempt to answer these questions once we have had a chance to look at the scale of the universe – and how far we can scale both upwards and downwards to even come close to matching the full scope of the Universe.

Chapter 6

The Scale of Physical Reality

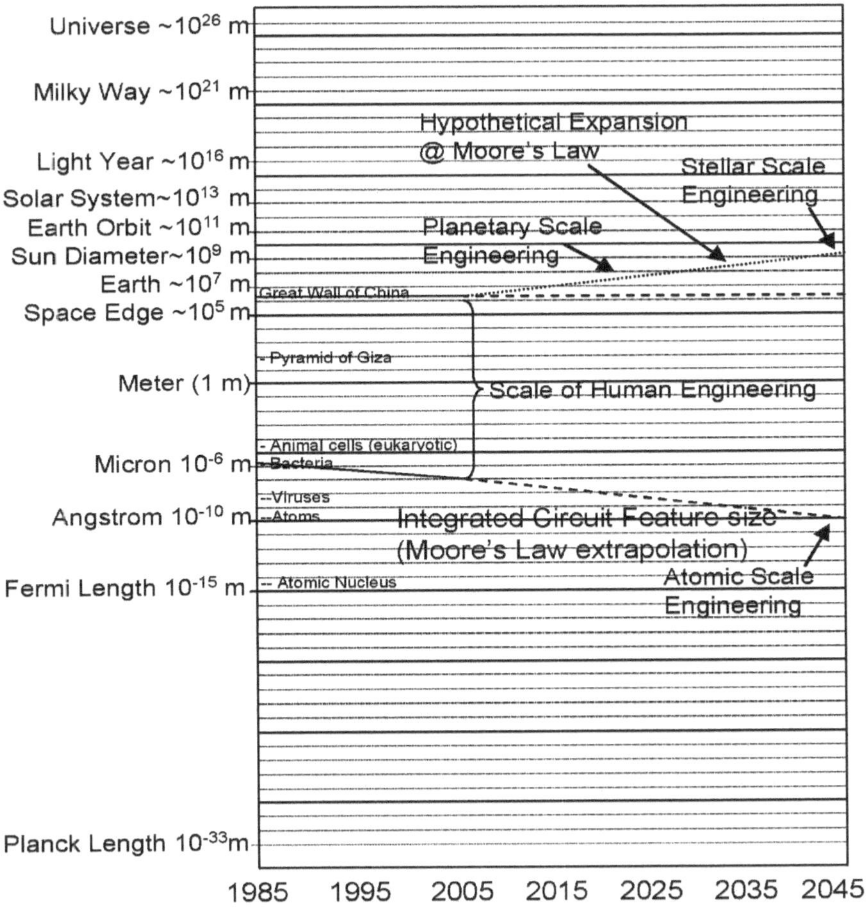

Universe ~10^{26} m

Milky Way ~10^{21} m

Hypothetical Expansion
@ Moore's Law
Stellar Scale
Engineering

Light Year ~10^{16} m
Solar System~10^{13} m
Earth Orbit ~10^{11} m
Sun Diameter~10^9 m
Earth ~10^7 m
Space Edge ~10^5 m

Planetary Scale
Engineering

Great Wall of China

Pyramid of Giza

Meter (1 m)

Scale of Human Engineering

Animal cells (eukaryotic)
Micron 10^{-6} m — Bacteria

Viruses
Angstrom 10^{-10} m — Atoms

Integrated Circuit Feature size
(Moore's Law extrapolation)
Atomic Scale
Engineering

Fermi Length 10^{-15} m — Atomic Nucleus

Planck Length 10^{-33}m

1985   1995   2005   2015   2025   2035   2045

Figure 6.9: From Planck length of size of Universe – the Human Footprint

A bit of explanation is necessary as we look at Figure 6.9. The vertical axis is logarithmic, and each subdivision represents a 10x linear increase or decrease. The smallest unit is the Planck length, below which the dimension of distance cannot be meaningfully subdivided (according to Quantum Mechanics). The largest dimension we have is the size of the visible universe. Expressed in Plank Lengths, the dimension of the universe is approximately 2 followed by fifty-nine 0's.

The Human Footprint

Against this backdrop, let us superpose the engineering footprint of Humans – i.e. how small or how large a quantity can we reliably engineer. If we look at the high end of Human achievement, we have the great civil engineering projects that can stretch across continents. Although it was not built in a contiguous fashion, one of the greatest acts of engineering is actually about two thousand years old – namely the Great Wall of China. This is not to say that intercontinental railroads and roadway systems are any smaller, just that they are of about the same scale as the Great Wall. The horizontal axis is a window in time, starting from 1985, and stretching forward to 2045. Hence, the top end of human ingenuity has been relatively flat for more than a thousand years, and indications are – it is going to stay that way for a while (dashed line).

At the bottom end we see a lot more activity. The semiconductor industry has been blazing a trail of miniaturization for the last few decades, and the dashed line projects the same rate of future engineering progress. To imagine just how significant this rate of progress has been, if we applied the same rate of feature scaling (in reverse) to the Great Wall – we would be engineering objects the size of the Sun in just 40 years! Today we turn out millions of Integrated Circuits with functioning units smaller in size than the cells in our body! If we keep going at this rate, within 40 years we will be able to build up matter one atom at a time – giving us the ability to synthesize any material or organism from scratch! Let us call this Atomic Scale Engineering.

Hypothetically speaking, if we were to somehow expand the top end of human engineering at the same geometrical rate that we are pushing into the microscopic side, we would be very close to being able to build a whole planet – with each atom exactly where we want it to be – in abut 40 years time. In addition we could populate it with organisms with very precise genetic characteristics. This does not address the moral issues of creating life that may even be superior to our current existence – but is intended to show that it is well within the realm of possibility. In spite of all the evidence to the contrary – we need to remember

that a sufficiently powerful God could indeed have created us and our entire surroundings with relative ease. And yes, if we persist on our technological march forward, the question will not be – whether we could do the same ourselves someday? The question will be – would we have the moral authority and philosophical fortitude to undertake such an endeavor? What if the progeny civilization reaches a potential that might be far greater than our own? Or the converse – will we be prepared for the possibility where things go terribly bad (e.g. Nuclear Annihilation)? Again, moral implications apart, would it take at least a Supra level Entity with the intelligence of 10^{14} humans to undertake such a project? In the author's mind – there is no doubt that the answer is yes – and it is very likely that a similar set of challenges exist for every subsequent level of encapsulation that human beings might participate in.

<u>Supra Levels</u>

In a three dimensional world, each linear 10x expansion actually translates to a 1000x increase in packing density. Although humans and the cells that we are constructed from are only 5 bars apart in Figure 6.9, it translates to a 10^{15} cells in the Human organism – if all we were was solidly packed cells. We do know, of course, that there are a lot of fluids and structural overhead, so a 10^{14} cell estimate in the Human body is quite reasonable. So, without any new technology for miniaturization beyond what biology has already achieved, a supra level entity with the capability of 10^{14} humans could fit into the 3D space of approximate dimension 100 Km. As figure 6.9 illustrates, when we are able to get down to an atomic level of engineering – we can expect the physical size of a Supra to be considerably smaller.

The 5 bar difference also happens to be marked with a solid line in Figure 6.9. So, at least in theory, 10^{14} Supra entities could decide to go Super Supra (S2) with a little bit of elbow room in a body slightly bigger than the dimension of – the Sun. The universe does not run out of space after that – the process could be repeated three more times before we ran into the edges of the Universe! Yes, the Universe is large indeed – and we are barely beginning to scratch the surface of what is possible!

From the viewpoint of the sanyasi, we can begin to see the scale
of the opportunity that we have to grapple with - in coming up
with the grand vision of universal success that not only spans
all humans, but can be extended to other intelligences as well.
It is for this reason that it is necessary to release the sanyasi
from the travails of day to day existence and the continuous
functioning of the Vanaprastha Network. No longer will the small
rewards of setting and achieving relatively short term goals be the
mechanism to keep the sanyasi internally motivated. The internal
joy to persevere on the path to greatest achievement must now
come from Ananda.

Ananda

It is a characteristic of human beings in the first three stages of
existence - we want to see measurable progress in return for our
efforts – in a reasonable time. If we do not, we get disheartened
and may decide to look for other avenues for our efforts that
could be more rewarding. The sanyasi does not have the luxury
of waiting, and breaks this cycle of self fulfillment by setting and
achieving goals – through Ananda or Spiritual Bliss.

We understand the time bound desire and gratification process
in ourselves as the Ego, as we discussed in Chapter 4. The
Ego does not go away at the Vanaprastha stage – it just gets
transferred to issues that are bigger than our selves and our
families. The Ego can be thought of as a way of projecting the
personally 'desirable' cause and effect scenarios, and expecting
that over time we will see our hoped for scenarios come to pass.

During the Sanyas stage a marvelous transition happens – the
Ego becomes timeless – and the cause and effect dependency
that originally housed it becomes quite meaningless. The fountain
of purpose and self worth that was previously fed by the process
of setting up a series of cause and effect relationships throughout
our life is now tended from the vast ocean of Ananda. In Sanyas
- "To be or not to be? "– is not the question any more. Our life
and our passing away are not polar extremes – they are but two
aspects of the same existence.

Ananda is the joyous sum total of the good we have done in our lifetime, what others have done before us, and the unimaginable accomplishments that are to be had in the future. In the process of defining our contributions to the Cultural Genome during the Sanyas stage, it is expected that most of us will accentuate the positives and significantly eliminate the negatives. We know from practical experience that all negatives cannot be eliminated – as the power of the positives can only be vetted against a backdrop of the negatives. Just as the speck of sand that irritates the oyster into creating the lovely pearl is forever embedded within the beauty of the pearl; similarly it will be necessary to capture and encapsulate the negatives (or irritants) that lead to the creation of all that is good around us. In net, it is expected that the summation of the human experience as of today, as well as our future accomplishments undertaken under the umbrella of Yogic Unity, will be overwhelmingly positive. Hence the state of Ananda is full of joy and hope in the good that has been done already - and in the belief that we are participants in a timeless experience that will be joyous and rewarding beyond imagination.

A special mention needs to be made here relating Beauty to Ananda. Yes, beauty is in the eye of the beholder, but there are exemplary works of creativity like Michelangelo's David and Shakespeare's Hamlet that have a permanent place in our Human Cultural Heritage. How do we define something that we would consider beautiful? Is Einstein's sublime deduction $E=mc^2$ beautiful? Is Copernicus's vastly simpler model of the heavens, where all the planets rotate around the Sun, beautiful? Beauty is often considered to be sublime – i.e. capturing just the essence of something without all the distractions and imperfections. We know that beauty of the flesh is skin deep and impermanent – yet with MAP even that can be given a permanent form.

In the immortal words of John Keats "A thing of beauty is a joy forever". Beauty has to do with simplicity ($E=mc^2$). It has to do with a high level of functionality – like a sleek sports car. It has to do with fulfilling the yearning in ourselves for an accurate and complete representation of something very familiar. Beauty reminds us of our quest for perfection, which though elusive, gives us great pleasure in getting close to it. The Taj Mahal is beautiful,

although the love story it embodies passed away centuries ago. A person caught up in the bustle of a time bound existence does not usually have the time to truly soak in the beauty around us. The sanyasi – in being able to disassociate their self from the haste of a time bound existence – can take the time to discern the hidden beauty that lies around us. The sanyasi can then take the most sublime works of art and wisdom and preserve them as a beautiful Gem for posterity. Ananda and the appreciation of Eternal beauty are but two sides of the same coin.

Sanyas and Fatalism

As we are still in the pre formulation stages of the Human Cultural Genome – it is expected that what we see around us is raw and unrefined information. Let us put ourselves in late 2004, with the great Asian Tsunami having just happened with a total fatality number inching close to the quarter million mark. Technologically, we have not yet been able to put a sensory envelope around our planet to even know when such disasters might be coming our way. To a lot of folks that have experienced the destruction and unhappiness around us it is easy to see why there are so many of us who might believe – "Life's a Bitch, and then you die!" For a lot of people – instead of perceiving the future of humanity as a series of glorious accomplishments and the fountainhead for Ananda – there is a profound lack of hope. The state of resignation to an unattractive life followed by death or destruction is referred to as Fatalism. Fatalism is the antithesis of Ananda – and is marked by a profound lack of hope for a better future.

We have discussed the concept of Ananda – and seen how the truly spiritual do not have to rely on drugs or the satisfaction of desirable outcomes to be happy. The spiritual joy comes from the well spring of all that is productive and joyful, ever to be associated with our human condition – and is truly timeless. Can too much Ananda be a bad thing?

An argument can be made that the concept of Ananda is unproductive, and not conducive to achieving any significant goals in life. If joyfulness can be had for free, who would want to work? If the well spring of joy is indefinitely large, it should not matter if

we add to it or subtract from it. So what would be the motivation to do anything to add our thimble full of contributions to this immense ocean of happiness? This logic too is another form of Fatalism – suggesting that we cannot do anything to improve our destiny.

At this point we should note - the stage of spiritual development where we can directly tap into Ananda is a relatively rare spiritual accomplishment. Ananda is associated with Sanyas, not only because the sanyasi has paid his dues and now can be rewarded with access to eternal bliss. It is also because the psyche of the sanyasi, having had a good grounding of life here on earth - can deal with the infinities and eternities surrounding our condition. An accomplished sanyasi is indeed an exalted spirit, and continues with their efforts even if there are no acknowledgements or rewards forthcoming.

As a percentage of the total population, the number of such enlightened souls is extremely low today. The expectation is, with the help of technology, the percentage of Human effort it takes just to sustain us economically will continue to shrink. What would humanity do with all these spare cycles? "A mind is a terrible thing to waste" is the slogan of the United Negro College Fund. By concentrating on Social Service the Vanaprastha will be able to turn otherwise idle cycles into a unifying Social Network. Similarly, by working on timeless Spirituality, the sanyasi is able to turn his or her energy and perspective into a greater purpose for Humanity! Without Ananda, there is no pot of gold at the end of the rainbow – and little incentive for the successful vanaprasthi to move into Sanyas.

The deeply religious amongst us might believe that the best part of life starts with our passing away (i.e. upon union with one's God) – and hence is not connected to our current physical existence. These people still have Hope, even if it relates to a different plane of existence – and with Hope comes Ananda. So whether or not we believe in a religious doctrine or a scientifically possible Glorious future – the real difference between the two outlooks (Ananda vs. Fatalism) is the presence of Hope as part of a beautiful and satisfying totality of human experience.

From Zero to Infinity

In the Mahashrama Cycle, we start from a very small existence (say the DNA and cellular processes of a sperm and an egg), grow through the Brahmacharya and Grihastha stages, and then slowly shrink in economic footprint down through our Vanaprastha and Sanyas stages, as we get ready to pass away. We start off very small, and when we die, our life's experiences probably can be compressed down to a set of digital content (our Cultural DNA) comparable, in size, to the biological genetic content we started off with. So we do start small, and end small – not quite zero to zero – but something very close.

That difference from zero is very important, however. The genetic material we start with is very close to the billions of other people that also inhabit this earth. The life's experience we accumulate and crystallize key learnings from – are quite unique to us. If we do not pass on this collective wisdom, humankind is forever the poorer for the loss. Collectively, all these near-zero contributions add up – and multiplied across the collective human experience can be large and significant enough to change the very nature of the Universe around us. The sanyasi believes in the powers of indefinitely large amplification and is willing to invest his or her talents in crystallizing these gems of wisdom - and then propagating them in a way that retains their utility over large tracts of space and time. Over time, these contributions build up, and can take us to a whole new level of existence.

Once we achieve the next level of existence, the accumulation of Cultural DNA does not stop there. As we have seen, the scale of the universe is such that the relative scaling from human to Supra can be repeated four times over before we run into the edges of visible space. At each level the metamorphosis to the next level happens through the accumulation of Cultural DNA, and by the sanyasi figuring out what existence might look like for the exalted entity. The role of the sanyasi, therefore, is critical to the success of the entire population as we seek to transcend our existence into something that is much more rewarding and immortal.

Chapter 7
Immortality: Preservation and Reenactment

Immortality is the genius to move others long after you yourself have stopped moving
-        *Frank Rooney*

Stepping away from the Grand Vision of Sanyas – let us now take the opportunity to examine what it really means to preserve the essence of being Human. We will look at preservation from the point of view of the individual, the group (e.g. ethnic entities) and from the perspective of the entire Human Civilization. In addition to the art of preservation, we will also examine the process of re-enactment or recreation – not only from a perspective of what actually happened (Historical) but also from the point of view of what might happen (Future).

In working through the various uses of preservation, re-enactment and simulation, we will discuss some of the alternate scenarios of human primacy under which the accomplished sanyasi might have to use these powerful tools. As in any discussion that has to do with Sanyas – please be prepared that several scenarios will be presented that would stretch our very concept of reality. Hopefully, this would lead us to an appreciation of the tremendous task that the accomplished sanyasi is up against - as they try and create an all-inclusive vision, first for humanity; and then for all intelligences in this universe.

The art of Preservation

Preservation is all about storing our Cultural DNA for eternity – or as close to it as is consistent with the Universe's own life cycle. As our ability to store more and different kinds of information increases – to some extent, so does our definition of what we consider central to our identity. For example, we may consider our finger prints as part of our core identity today, but not our toe prints. Over time, more and more of what defines us will become 'core' – especially as more lifelike reenactments become possible.

In the extreme, the information on every cell and every hair of our bodies could be preserved – which if it were reconstructed in the future might indeed make us feel like we have woken up from a very deep slumber.

The march of technology as it relates to storage capability is indeed remarkable. We are rapidly getting to the point where there is sufficient permanent digital media (e.g. Hard Disks, magnetic tape) available to write the stories of every single person on this earth. In a few years we can not only capture the text (i.e. that which can be put into words), but much of the context as well, the picture, the smell, the color, the texture of the skin of the newborn infant as you hold her for the first time ...

We have noted in Chapter 5 the very fast exponential clip that storage densities are on today. We can expect that by about the year 2030 storage capacity for both text and context will become so abundant it should be possible to record every single waking moment of every single person on this earth - should we so desire. Let's pause for a moment and take a look at where all this content comes from. Some of it could come from sensors embedded onto our clothes, or accessories – like ear-rings. There would be several of these sensors collecting all kinds of sensory and biometric data – and this would be the on-body network. In addition we could co-opt the inputs from locally available monitors and sensors that would give a third person view of ourselves. This would be our out-of-body network. People working together could decide to share their sensor resources – and then later pick out the ones that they think are worth preserving.

However, all context is not sensory, or even external to our own minds. The state of mind of the individual is also a real part of our context. No technology is yet available that can look inside and accurately follow our thoughts – but research shows a lot of promise. Significant advances have been made in figuring out from our brain scans if we are happy or sad and even the color we perceive with our eyes. With some training people are already able to move around a cursor on a screen with only the help of a non-invasive thinking cap. Humans have a highly asymmetric relation between the amount of data we can track through our eyes and ears, and the amount of data we can output – say by

typing on a keyboard or even by speaking. Hence, sensing the mental context in a replay session will be much easier than the original process of encoding it. Capturing mental states and thinking processes will take considerable mental discipline and objectivity on the part of the sanyasi putting together a reenactable Cultural DNA.

There is no need to edit our life experiences if we do not want to – but once in the Sanyas stage we might decide to go back and do some editing, and add some commentary – much like a Director would edit a movie in the movie studios. Some experience may be just too painful to remember, or appeal to such a gross level of our existence that it is left out on the cutting floor. If we leave out too much – we run the risk of missing key aspects of what made us the people that we are – and risk coming through as too superficial. If we include inconsequential detail (like the make and model of every car that we passed on the way to work each day) – then we risk coming through as an existential pack-rat. The mental energy that we put in to separate the wheat from the chaff is what makes our encapsulation of our Cultural DNA look like a 'Thing of Beauty' that should be preserved 'Forever'.

This is also the time to look through and see if we can really condense some powerful insights (Gems), and offer it up as learning for posterity. For this we need to look at what constitutes 'real' staying power. There are a very few things that have the staying power to last – and possibly remain relevant – for a very long time. Examples that we have used to illustrate the Sanyas beat rate include the US Constitution, the Bible, the Koran and the Bhagavat Gita. Not a lot of our individual experiences would have this kind of staying power or Gem quality. It is the prerogative of the sanyasi to offer up the best Gems of inspiration for the rest of us – even if humanity has to wait for a future scholar to come and string them together into a stunning necklace that becomes an enduring icon for all of Humanity.

The part of our cumulative experiences, deductions and thought patterns that survives the director's cut of the sanyasi becomes our Cultural DNA – which we would offer up for posterity. The task gets a little harder when there are multiple people involved in collecting and collating the Cultural DNA of a group or ethnicity.

Here, it is no longer sufficient to trust just one person's judgment as to what is relevant and what is not – and what remains after all the edits has to make sense to the entire group being represented. The task gets even harder as we try and develop a unified Cultural Genome that is representative of the entire Human existence. Yes, some aspects of MAP need to be practiced even in the codification process – so that we can represent the whole as more than just the sum of the individual perspectives.

Simulations - Forward and Backward looking

We have discussed in the Vanaprastha section how individuals, ethnicities or even nations in conflict can use the tools of MAP to figure out a solution space where the totality of the cultural interaction is greater than the sum of the individual parts. We have also noted that the best way to come to a consensus on moving forward is to formulate various alternatives, and then test them out using cultural simulations to see which one(s) are the most likely to succeed. This is a forward looking simulation, which may converge on several eventualities – each of which must be carefully considered in coming up with a weighted evaluation for a particular path of action.

In a backward looking simulation we have the benefit of historical hindsight, and we can fine tune the simulation to get us to a known eventuality. Yet, at every step that we have verifiable historical information, there will also be some contextual information that will need to be deduced - and a high probability path has to be proposed that explains all the observed facts. Thus, even with the benefit of 20/20 hindsight, some amount of alternative scenarios do need to be considered and evaluated to come up with a free flowing replay of past events and accompanying thinking patterns by the players involved.

Replaying Reality

Supposing we are future students of history, what would be the ideal situation for studying a perfectly saved, high-fidelity, community or personal perspective? What kind of maximally

immersive experience would we re-create so that our future student could really sense and understand the life's lessons of the person or culture being studied?

We have introduced the concept of Virtual Presence earlier in this book (Vanaprastha section). Let us also note the similarity with what electronic gamers are doing already with their highly immersive games. When we try to replay reality we need Virtual Presence with an enhancement – now we not only need the ability to soak in their context and react to the virtual environment – but also sense the very thinking process and cultural nuances that are in play.

If our children are any indication – we can extrapolate that humans will get very good at the Virtual Presence skill set. Already there are simulated reality games where the individual player controls not only the actions of a single person or vehicle in a virtual game playing environment, but several. We see a lot of eye tracking and even brain wave tracking work going on with how the fighter pilots today interface with an incredible amount of information about their craft and their surroundings – and in some cases just have to 'think it' to fire their weapons.

Moving forward another 40 years or so – the electronics and flight systems under the influence of Moore's law could be so miniaturized that a single pilot can control hundreds or even thousands of Hornet size remote controlled fighter aircraft – and must come up with a way of keeping them all coordinated. The 'hornet' master of the future will need to coordinate multiple hornets simultaneously, learn to give each an appropriate level of autonomy, and will need a significantly more sophisticated and immersive interface system than even today's fighter pilots. A remarkable thing is expected to happen when the human brain is trained to simultaneously support a variety of interfaced devices. Some of the thinking process itself can now be outsourced to machines. The thinking processes may be how fast and how high to fly the 'hornet' swarm in order to avoid detection – or how to fill in the ranks after part of the swarm is incapacitated. And yes, there must be ways to overrule the outsourced thinking processes with what we call 'gut feel' or intuition. The remarkable thing is – such mental dexterity fostered by immersive Virtual Presence

technologies can now be used to sense human thinking process in a simulated environment - just as if it were another set of game inputs that came to us through the machine interfaces.

This extended Virtual Presence concept can now be applied as we step into the shoes of persons long passed away - to be able to view the world from their unique perspective. This study is not a pen and paper decoding, and many dusty hours of research in a library or an archeological dig to painfully reconstruct what life was like – say to be an Anasazi Indian. Now we can speak to our host, and even frame ourselves inside his or her mind; to walk in his or her footsteps and really find out what life was like to be them! It is quite possible that in a hundred years, we will no longer just open up a family album to show the children what Great Grandma looked like. We will take the children along with us in a fully simulated 3D trip where the children actually get to talk to Great Grandma and hear stories about how things were during her own childhood – or even experience the happenings for themselves.

An accurate rendition of the condition of a person that is replay-able has value even if the person is still alive. We have already determined that a major part of conflict resolution is putting out an encapsulation of the core identity and core interests of the negotiating parties (cultural Genome) before we even start a good faith negotiation. Having experienced each other's perspectives, troubles and challenges, what are the chances that we cannot come to an effective compromise? Yes, peace may even be possible in the Middle East.

The Vanishing Viewpoint

There are many sub-cultures, languages and lifestyles dying out today as mankind rushes headlong into the arms of globalization. Even for the author personally, the mastery of the grammar, spellings and phonetic conjugations of his Mother tongue – Bengali – is rather spotty. Our children will probably be able to understand, but really not be able to converse well in their ancestral language. Their children subsequently will likely have very little connection to the original culture that we came from. The same scenarios are playing out whether one is a Navajo (Native

American tribe), an Armenian or an ethnic Fiji islander. The next hundred years will either take these cultures and make them immortal – or leave so little residue that an ethnic reconstruction becomes almost impossible! Hence, the time is ripe for the last adherents of ethnically 'Vanishing Viewpoints' to work urgently to get their ethnic cultural genomes transcribed in a faithful and simulate-able form. This would preserve their core essence of being for generations to come!

Testing the Cultural Genome

The preservation of the central aspects of existence for a specific host or host group - we have referred to as the creation of the Cultural Genome (or DNA). There is actually even a way to test how robust a Cultural Genome is, and how carefully it is recreated. If the virtual visitors (with the appropriate state of mental maturity) by and large follow the thought processes and come to the same set of decisions and actions as the original script (without prior knowledge of the outcomes), then we can say that we have a faithful preservation.

At this point, we do run into a chicken and egg situation. The encoding needs to be done with the technology available today – whereas faithful immersive simulation capabilities will keep improving over time. It is expected that some amount of archeological reconstruction will be needed – but hopefully in only the most minor details. The skill set required to do the cultural encoding in such a way as to be highly reenact-able is something that we will need to entrust the sanyasi amongst us to master and perfect over time.

Levels of Simulation and associated Overhead

Those of us that have worked in the business of designing integrated circuits (ICs) know that there are several levels of simulation – from fine grained to coarse grained. Fine grained (circuit level) simulations are usually run on small parts of the circuitry, and their behavior then encoded at the next level up. Even at a medium grained level (called the Register Transfer Level or RTL) – it takes a long time to simulate a full chip.

Also, since all the internal architectural detail is contained in the RTL simulation, and Chip developers are loathe to share this too widely! The next level of simulation modeling is called the behavioral level – where much of the internal workings are hidden, and what is expressed is how the design would behave (or interface) to the outside world. The behavioral simulations can either have a tight (cycle accurate) time response model – or a relative loose (bus functional only) response model where the fine grained time distinction is lost. In this example drawn from IC design, the details are not as important as noting that we can trade off accuracy and realism for speed and complexity.

	Speed (Approximate)	Complexity (Approximate)	Timeline (Availability)
Target System	1x	1x	Current
Hardware Emulation	20%	20x	Pre-Silicon, Hardware
Bus Functional Behavioral Sim.	10%	10x	System Validation
Cycle Accurate Behavioral Sim.	1%	20x	Sub System Interface
RTL Structural Simulation	0.1%	50x	Sub-System Design
Gate Level Simulation	0.01%	100x	Sub-System Design
Circuit Level Simulation	0.001%	1000x	Individual Circuit Design

Table 7.1: Levels of Simulation, and the associated slowdown of re-enactment

Working at a high level of abstraction allows relatively unsophisticated computers to simulate the working of a next generation CPU (central processing unit) – say at about 10% of the actual speed of the actual circuit. As a rough approximation, each level of finer granularity adds about another order of magnitude to the execution time, as well as increases the size and complexity of the system required to do the simulation

– as illustrated in Table 7.1. Hence, except for extremely small systems, it is impractical to do full system fine grained simulations in the typical IC (Integrated Circuit) design environment. The technology for simulation is a highly competitive endeavor, and the numbers given as examples are all very approximate – and intended only to illustrate the tradeoff between realism, recreation timeliness and complexity.

The dynamics of the game changes somewhat when the task at hand is re-enactment rather than chip design. With superior technology a 1,000,000x increase in complexity and a 1000x speedup is definitely possible over time. There is no magic in these specific numbers – i.e. a larger complexity increase could make up for some level of lost speed. If this can be done with a general purpose computer – it represents approximately a three orders of magnitude linear compression – or about 40 years of wait time at the current geometric rate of Integrated Circuit miniaturization (also called Moore's law). If on the other hand, we use a special purpose emulator to do this – the wait time comes down dramatically. Now it becomes possible to build an emulation unit with the same complexity and even faster internal processing in just 9 years! Even then, there will always be an availability gap between the real life system being simulated, and when a structurally accurate simulation or emulation can be equally responsive. The people who will be tasked with taking an old cultural genome and fleshing it out to where response times come close to real time – we will refer to as Cultural Archeologists.

It can be extrapolated that Cultural Archeologists armed with future technology can come asymptotically close to the original event line when recreating history. In the process they will find many decision paths that were Historical dead-ends – but in the realm of quantum probability could very easily have been taken. In fact, even with the most detailed History, there will be gaps in the timeline that the Cultural Archeologists will need to develop alternative threads for - until one of them is known to have been taken.

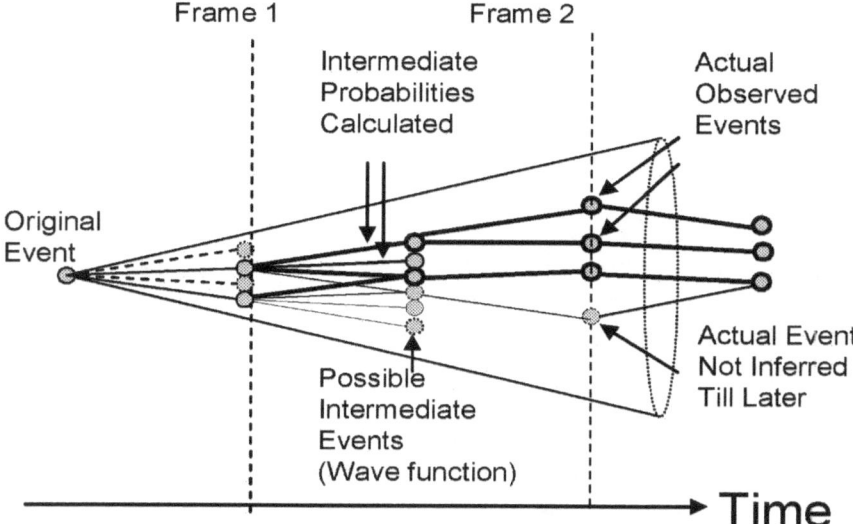

Figure 7.1: Bridging between the known states – the need for quantum computing

In the wonderful world of Quantum Mechanics, it is perfectly acceptable for a particle (or a machine) to be in multiple simultaneous states. Even in the human brain, if we come across a new word during the course of a conversation, we will mentally keep testing it against several possible meanings until one of them wins out with additional contextual information. What the Human consciousness tends to do is to bring down the multiple possible renditions of what could have happened to a single historical thread of what (we think) actually happened. This is often referred to in Quantum Mechanical circles as collapsing the Quantum 'wave function'.

In Figure 7.1, we look at a set of events along a time line, and superimpose on it a frame-rate during which the events are observed. As you might already know, all multimedia streams, especially video, is comprised of discrete frames – which when played in quick succession gives us the illusion of continuous motion. Similarly when any set of events is recorded for posterity, there can be assumed to be a discrete frame rate. In between two consecutive recorded frames, we don't know precisely what happened. Also, even when an event happens, there may not be

an immediate visible effect – and the fact that it happened has to be inferred from future observed events. What this points to is the need to be able to carry the probabilistic wave function forward until a set of observed events collapses it down into the annals of recorded history.

It is a widely held belief today that the next big jump forward in computing power will come when we really master Quantum Computing. By its very nature Quantum Computing is especially adept at simulating multiple probabilities simultaneously. This will be a major boon to the Cultural Archeologist as they try to re-create a detailed functioning simulation based on our Cultural Genome. This will also be a major benefit to the sanyasi of the future as they not only formulate our Cultural DNA, but also test it to make sure that the possibilities developed end up explaining the observed occurrences.

With adequate training the visionary sanyasi does not need to collapse the Quantum wave function! All other individuals that are caught up in the day to day cause and effect relationships must choose a single thread for their existence. The visionary sanyasi can train their minds to see existence in all its multiple probabilities. They can simultaneously work at reaching the best existence for Humans assuming more than one outcome holds true. Let us do a quick mental exercise to illustrate how you, the reader, might stretch your mind to create a vision of success in the presence of multiple end games.

Let's go back to Figure 6.8 and put ourselves in the shoes of a visionary sanyasi trying to figure out the best future to be had for humanity. Let's say that each quadrant in this figure represents an equal probability of 25%. Looking at the chart, we can calculate that the chances are greater than 50% that the right human response would be to build up our capabilities through Yogic Unification. Would this be the way the visionary sanyasi reads this chart? In integrating through the various probabilities, the sanyasi will see that the starting point is most likely in the Brahmacharya segment, even if the ultimate end-points end up being distributed equally between the segments. The discourse has to start with Individualistic seeking, and hopefully arrive at a small set of eventualities for human primacy that most humans can agree on.

Through the use of MAP, the protagonists from each camp would take a virtual tour of the competing perspectives, and see if they could come up with a unifying vision. Through the process of Yogic Unification – if most of humanity can arrive at a subscription of one or two common threads – work can begin in earnest on their implementation. This may seem like a roundabout way of getting to the same end-goal, but the inclusiveness of the process makes it much more likely that most of humanity would be on board by the time it takes to implement the vision. In the absence of cultural simulation capability, the visionary sanyasi would train hard to use his or her imagination, to work through the multiple paths that lead to the highest probability of success - for the grandest realizable vision for humanity. This is multi-dimensional, time integrated, multi-eventuality optimization, and a fitting challenge for the advanced sanyasi.

Coming back to the topic of preserving for future re-enactment throught simulation, we still have the minor inconvenience of not having a powerful enough set of tools to test our just-preserved database. In the time frame that we are waiting for a realistic re-enactment technology to arrive – it will be up to the visionary sanyasi to use their own mental skills to fill in adequate details for a preserved set of events (and viewpoints) to be simulate-able and recreate-able in the future. Ideally, the lag time from original preservation to a full fledged real-time simulation capability will not be so long that the sanyasi responsible for the original database is no longer around to help with the Cultural Archeology of faultless reconstruction.

The difference between a historically accurate replay and a realistic simulation is worth noting in the context of dealing with multiple probabilities. In an accurate replay, the quantum or probabilistic wave collapses with every frame of captured data, and a high correlation is maintained to historical observations at all times. In a simulation, the intermediate probabilities are not thus constrained – and might indeed carry on for considerable periods of time. Hence, in a simulation it is possible to have Sir Isaac Newton meet Professor Albert Einstein and have a discourse on the laws of gravity, as it relates to the orbit of the planet Pluto. As entertaining as such a session might be - it would never claim to be a historical reproduction.

Filling out the Cone of Causality

Just as goods manufactured today can have multiple uses with multiple customers, we have to be prepared that our preserved experiences will have multiple audiences throughout the future. It is also very possible that future entities will employ technologies that have so much computational resources that they decide not to collapse the wave function with every observation – but let it go on until every possible outcome has been exposed and investigated. This can be thought of as filling out the cone of causality – and learning not only from what actually happened – but what could have been. In the later twentieth century, the nuclear 'Doomsday Clock' came ever so close to midnight multiple times – but in each case total disaster was avoided. Is it not much better that we fill out the unhappy branches of the cone of causality through simulation than through actual destruction?

Not all threads in a filled out cone of causality that are filled in through simulations need to be the destructive end points. It is possible that a series of insights that could have happened early on in our human development might have been cut short due to unforeseen circumstances. Filling out the positive occurrences and developments might help us find Gems from a simpler place and time that could be used to lay the groundwork for the cultural DNA of a future entity. Our existence here on earth today may just as well be the exposition of one of the favorable threads that was missed in the original run up to a Prime Ascension – that our inner selves are but actors grafted in to a Category 4 Virtual Presence Gasket described later on in this Chapter. Probabilistically speaking, it is quite likely that we humans today are but players acting out the omitted threads in the cone of causality – leading us to ask – what happens when our particular thread has run its course and produced all the gems worth harvesting?

There are two end games likely in this scenario. One is – the most favorable threads are put on an accelerated timeline and they ultimately catch up with the actual timeline. At that point, it is as if two parallel civilizations merge to create one that takes in the best of both. Theoretically this need not be limited to just two threads – we could have several that merge into the original at different

time intervals. The second end game is – our thread is not deemed to be favorable enough, the relatively sparse gemstones are harvested, the experiment halted, and the actors set free to return to their original timeline. If the difference between the two end games does not give us enough of an incentive to make our particular thread the best that it could be – we would be hard pressed to find a better reason!

There is a third end game as well – which speaks to a possible deficiency in the evolved lineage of a superior intelligence. It is possible that what is being harvested is not just the gems of wisdom, but the 'New Blood' in the way of Mahashrama tested individuals that are raring to go to build a bigger 'Vision for Intelligence in this Universe'. Great minds, in their old age, tend to lose their vitality – and that is possible of highly evolved parent intelligences as well. The positive side of this scenario is that the New Blood is now able to find an even greater pedestal from which to launch their Universe wide strategies. The not so positive side here is - the resistance is too great in the highly evolved intelligence to new ways of doing things. Now, instead of changing the new reality around them, the New Blood effectively becomes a source of entertainment and education, and a way to live vicariously in a livelier and happening timeline.

The sanyasi working on creating the cultural genome must keep these scenarios in mind as we try to carve out the greatest possible future for humanity. The base scenario, of course, is that ours is the first intelligence to have come thus far, and need to plan to have our human cultural and philosophical DNA to have universe wide acceptance – with no assistance from a higher intelligence. Yet, all the other scenarios mentioned here also need to be taken into account, and possibly a few more. To anybody but the accomplished sanyasi working in probability space, such an open ended scenario list might stop us dead on our tracks – or even send us into fits of consumptive frenzy because "who knows what tomorrow will bring?" To the sanyasi, grappling already with the eternal and the infinite, these are just a few more wrinkles to deal with in probability space!

The sub-nuclear domain, and the Elasticity of Time

If we go back to Figure 6.9, we see that there is a considerable amount of relative space (18 orders of magnitude) between the Planck Length and the Fermi Length. This is the sub-nuclear space, and associated with large amounts of matter and energy interacting within tight confines. Much of our knowledge of the sub-nuclear domain comes from accelerating sub-atomic particles to almost light speeds, and causing high energy collisions between them (or with ordinary matter). Human knowledge of this space is still very rudimentary. The places where conditions support this type of interactions are: inside the denser stars and black holes – or during cataclysmic events like the Big Bang or the possible Big Crunch. If quantum computers were to be built that really took advantage of such conditions – it is conceivable that several human lifetimes could be simulated within a relative 'split second'. Far in the future when we have access to such computationally powerful environments, time becomes very elastic. It becomes possible to create accelerated timelines for civilizations with high potential, and really slow down and even put on suspended animation those others that might turn out to be destructive or unfulfilling.

A Special Place for Humans?

In a typical simulation, certain things are held as constants (e.g. power supply voltage on an IC simulation, Laws of Physics in our Universe) – and certain others are variables (e.g. the date and time). We humans perceive an uncommon degree of constancy in our Universe. The fact that our telescopes can see all the way to almost the beginning of time – adds to the wonder of the linearity of space and time that we perceive. With the possible exception of a massive star or a stabilized black hole with a regulated energy flow, the other high energy environments that we talked about are wrought with overwhelming variability - with very few constants! Again, looking at Figure 6.9 – we look like a small number of germs that have alighted in the middle of a very large Petri dish. It is almost as if – it is the ability of Humans to scale up an appropriate answer to the magnitude of the Universe that is the subject of today's experiment. Yes, all of us need to accept

the possibility that we have been plugged into a Grand Simulation Experiment – like actors playing the story of how intelligence might develop in a wide open and relatively unconstrained environment.

Figure 7.2 shows the environments that exist in the Universe using two different metrics. Along the horizontal axis, we observed from Figure 6.9 – the scale of our existence (Human Engineering) puts us at very close to the center of what is possible. The other axis is the energy density – which requires a bit of explanation. In one extreme is the inside of neutron stars or Black Holes where a single drop has more mass than a supertanker on earth. On the other extreme is the vastness of open space – where one is lucky to find a single Hydrogen atom in a cubic meter! We already know from Einstein's $E=mc^2$ that energy and mass are equivalent. Thus, if we were to chart the scale of most human engineering along the vertical 'Density' scale – it would also be close to the center of the Energy Density continuum.

Given a certain amount of mass (energy) the preferred environment for perpetual storage is a very large universe with an extremely low energy density. In the other extreme, the best environment for processing happens to be tightly bound matter and energy that interacts strongly. To illustrate this dichotomy, even today's computer storage elements (e.g. Hard Disks) do not need power just to store information – but only to write and retrieve it. On the other hand, most modern CPU's tend to run very hot and need a constant supply of energy to keep computing.

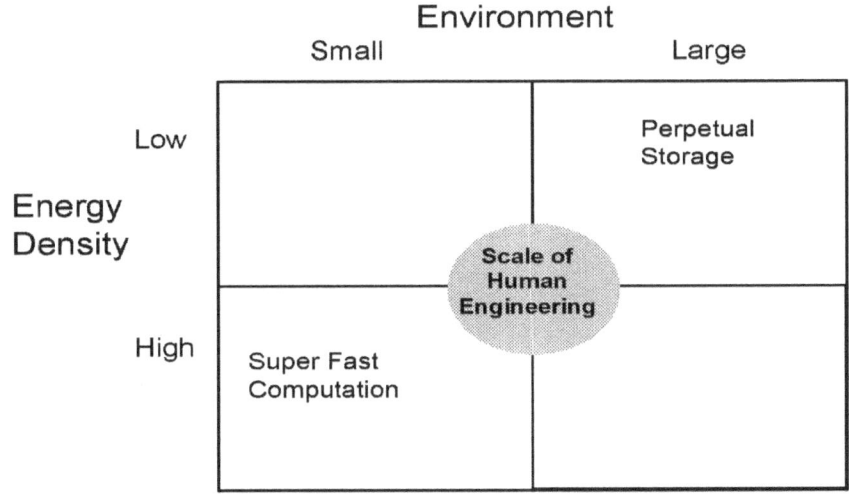

Figure 7.2: Best environments for long term storage vs. computation

If the kind of intensive simulations that would be needed to fill out a cone of causality were indeed to be carried out at the high energy sub atomic level – a mechanism must exist to take the useful output and store it in an efficient and long lived manner. This is just one more example of how the polar opposites of the Universe have a Yogic need for each other – and Humans just happen to find themselves right at the center of it all! Let us, therefore, keep in mind that all of this may be by chance; or by design – as a greater purpose in the Universe seeks to fill out the cone of causality of how they might have originally developed. A related possibility is that our thread is part of the universal citizenship rites of a civilization that has struggled to reach their potential because of extenuating factors – and feel the need to demonstrate that given a standardized starting point, they too would be capable of prospering under the umbrella of Yogic Universality. The next hundred years will likely determine whether our Human Cultural Genome has the potential to change our universe for the better. If successful, this could mean that our cultural and philosophical core DNA could be replicated many times – and work its way not only into the cultural DNA of

our Galaxy, but even our entire Universe. Under most of these scenarios this is a good time to live, and to build a vision for our potential as the Human civilization.

Cultural Simulations and Sentient Beings

When we look at the key players in any movie, the director takes special care to develop their characters and their motivations, as well as the conflicts and opportunities that arise. For the secondary players, however, this level of character building may be too much of a luxury, and they tend to be defined in relatively simple terms. Similarly, in a simulation it is important to develop all the key players in a way that is lifelike, and to equip them with the kind of internal consistencies and conflicts that defines a real human being. Let us take a quick look at some of the unique features that define us as human.

Consciousness needs no introduction – at least for us humans. Thus far, we do not have much of a clue as to how our brain really produces consciousness. We also have the subconscious level of existence – which can be tapped (say by a hypnotist) even if we are not consciously aware of our actions. Our consciousness, if properly conditioned, has the potential to plug into any of the four segments in the Mahashrama cycle – and is more central to our selves than even our core values.

We have already seen that the perspective we gain by going through the complete Mahashrama cycle allows us to relate to the individual segment needs for all segments. This puts the accomplished sanyasi in the unique position of being able to play any position in the simulation game! In fact, the involvement of such a person - who still has the energy and motivation to host a simulation or emulation – appears to be critical to any realistic Simulation or Emulation involving Humans. A parallel today is the talented Actor or Actress who can add life and dimension to almost any role that they would want to play – even if it were the role of an Ascetic!

A sanyasi who has successfully navigated all the four stages of the Mahashrama cycle is familiar with the Dharma (value) system that drives the individual segments, as well as the transitions

between them. Ancient Indian philosophy also held that our soul or consciousness would go through many rebirth cycles until we achieve the level of selfless purity that would allow us to achieve Moksha or Nirvana. In today's environment, where we have established that a single consciousness can navigate the full cycle within a single lifetime – we no longer need the concept of rebirth to mark our continued spiritual progress. One part of the game remains the same, however – the successful sanyasi who has circum-navigated the Mahashrama cycle is an entity with enormous powers to transcend our human condition.

Categories of Gasket Interface for Virtual Presence

For a consciousness to plug into a simulation – much like we would in any other Virtual Presence scenario – a series of interfaces (or Gaskets) can be defined. Figure 7.3 illustrates a series of circles that we can draw around our core consciousness, and the corresponding categories of Gaskets that would be needed. The largest gasket in Fig 7.3 includes all of our physical attributes (including senses and muscular capability). When we interface with the world on a regular basis – this is the interface that we use.

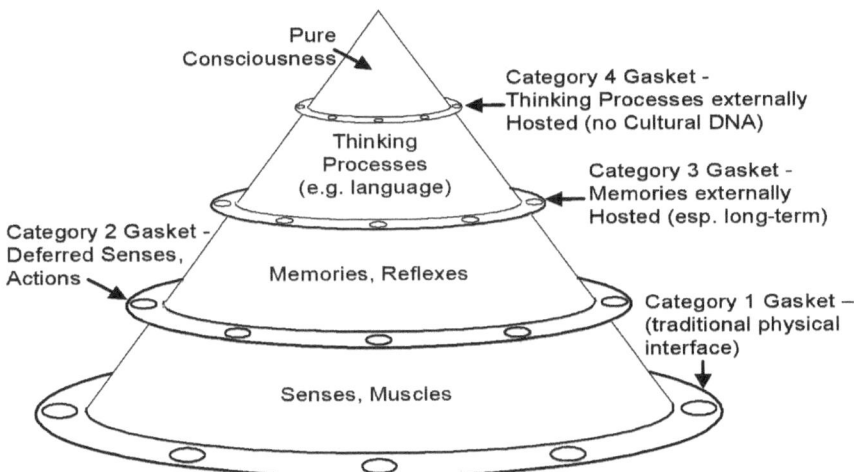

Figure 7.3: Categories of Gasket interface for Virtual Presence

The next inner circle leaves out part or whole of the sensors and actuators for a Category 2 gasket. To illustrate, the story came out in late 2004 about an enterprising safari operator in Africa, who had set up remote controlled hunting rifles that were target-able over the internet with the aid of surveillance cameras. The Virtual Presence game hunter would sign on and monitor wildlife over the internet, and if they decided to go for the kill – would receive their trophy via home delivery! Thus the hunter need never set foot in the same hemisphere as the big game, and would be able to experience the entire process via remote control. Although we might use our physical facilities to control a computer at this level – the interface could, conceptually, just as easily have been directly to our nerves that lead into our brain. This is an example of action happening far away from the human initiating the action. Another example of a Category 2 gasket is when the action happens at a different time – like an alarm clock waking us up in the morning. This would define a Category 2 gasket where we extend our senses (and actuation capabilities) to a place and time different from where we are physically located.

To illustrate the next Category, we look at people that have either temporarily or permanently lost their memory due to an illness or accident. In many cases they will still remember thinking skills they have learned (e.g. languages) – but the data (or content) seems to be missing. This suggests that, even within our brain, our memories are just a bit further from our core being than our thinking processes. Even in the case of computers, the applications (software programs) are closer to the core (operating system) than the data that is generated or consumed by these programs.

The Category 3 gasket is one that supports most or all of our memories within the machine part of the interface, and offloads our minds from having to remember the details. With the advent of writing, and now with the help of the multimedia computer and the internet – we already offload a lot of our rote memory tasks from our brains. At some point it should become possible to access and download information directly into our brains – but for now we will likely need to keep taking it in through our eyes. The reverse process of taking content out of our brain and into written

or computer form actually relies on a much slower mechanical process today (e.g. the typewriter/keyboard) and that too can benefit from a direct nerve connection. A tight man-machine connection often brings to mind the 'Cyborg' – who is half human and half machine. An invasive man-machine connection may be necessary when the sensory or action space is very demanding – either in response time, or in the amount of data processed. However, for most of us that are not operating at the cutting edge of responsiveness, just training ourselves to operate under a Category 3 gasket, and relying on machine based memories for most of what we need to do in our daily lives should come quite naturally.

Finally, we can conceptualize the case where we interface purely under the power of our consciousness and possibly a core value system (Dharma) corresponding to the Mahashrama segment we find ourselves in. It has been postulated that Consciousness is more a product of our emotional brain than our rational brain[1]. Hence, in this scenario, even the logical thinking processes are externally accessed. If we define a gasket around this core, even our thinking processes and our most fundamental human relationships (e.g. between a baby and her mother) would fall outside this circle. What we have just defined is a Category 4 gasket. This encapsulation then becomes truly portable – and able to bridge into entities with very different life circumstances – across generations of human (and maybe even non-human) existence. A sanyasi who has successfully navigated the entire Mahashrama cycle, and can detach their consciousness from the travails of day to day existence, may be seen as a pure consciousness in this scenario. In this case, there is no need to even carry a segment specific Dharma – because having mastered the Mahashrama cycle – the sanyasi has internalized the Dharma for all the four segments. As noted earlier, this makes them especially useful players within a cultural simulation as they can take on roles that integrate the entire human condition.

The observant reader might be able to draw a parallel between a Category 4 gasket and how we have defined Yogic Empathy in the Vanaprastha section. If we can devolve ourselves of everything outside the Category 4 gasket – we can plug ourselves into a

Cultural Simulation with an un-prejudiced perspective as to the case specifics. This is empathy in its purest form – where we do not have the distraction of our ego or our prejudices to cloud our judgment. Part of the training of the sanyasi is to try and devolve down to their very core personality – while still wrapped in the trappings of their physical existence. As each outer layer is cast out – what remains is a simpler and truer rendition of who we are at the core. This is also the process we would use to create our Cultural DNA – the core essence of who we are as humans, as communities, and ultimately as the human civilization. Thus, the sanyasi is not only the perfect actor to transplant into a cultural simulation – but is also the key architect to build the Cultural DNA on which a subsequent cultural re-creation can be based.

Read/Write Asymmetry in Virtual Presence Gaskets

In practical terms, the category of gasket interface could vary depending on whether we are reading from the environment or initiating an action sequence (writing). We have already noted the huge asymmetry between what we can take in through our eyes, and what we can put out – say with our fingers typing on the keyboard. At each level of gasket interface, there is a similar asymmetry as to our level of participation. Borrowing a metaphor from Shakespeare -

> *All the world is a stage, and all the men and women merely players.*
> *- William Shakespeare*

The interesting thing about a stage is – for every actor or actress, there can be hundreds of people in the audience following their every move. Even in daily life, there is a huge asymmetry between the amounts of information we take in; let's say through reading – and the quantities that we put out; let's say through writing. In most situations where we are interacting with a group of people, typically only one person will be speaking while others listen. When two (or more) people speak at the same time – we find it hard to follow either conversation. Similarly, if more than one writer tries to edit the same database simultaneously – we could end up with a corrupted database.

This natural asymmetry between the read and the write functions expresses itself in the gasket interfaces as well. Any output from the human mind – whether it is a thinking process (Category 4), memories (Category 3), space or time displaced action (Category 2) or just the simple action of signing our name on a check (Category 1) – needs to be clear and unambiguous as to who the author is. Once expressed, the written content can be accessed millions of times – as we do when we go to watch a movie, or even when we read a book. Hence, it is very likely that web accesses for various category levels of gasket interface will be granted in a highly disciplined manner for write accesses, and in a very liberal manner for read accesses.

The Evolution of the Web

As we have originally discussed in the Vanaprastha section, the Human Cultural Genome is essentially the aggregate Human context preserved in a way that is accurate, internally consistent to a large degree, and allows future re-enactment. It is expected that a lot of time and energy will be devoted by the sanyasi to developing the content and context in such a manner that it makes it possible to plug in a higher category gasket to any part of our cultural genome. This would allow a student of the future to re-live the experiences that continue to shape us as the human race. In addition to having meaning and context as an integral part of the web content, we can project yet another important step forward when we begin to see web locations that are simulation ready – paving the way for truly participative web browsing. A simulation ready web site could support gaskets all the way from level 1 to level 4 – with the challenges rising considerably for the higher levels of gasket hosting.

To the astute reader, the implications of this technology for entertainment purposes are probably already apparent. In fact the word that originally came to mind for re-enactment was recreation – which we have used minimally thus far in order to avoid the implication that this is all fun and games. The applications in education are also quite easy to grasp. It is very possible that our

future generations will find that so much of their daily needs can be met directly from the future web structure that they begin to spend most of their waking hours directly plugged into it!

Today, the way we interface with a computer is very similar to how a mother cares for a baby. We ensure it is powered up, has a connection to the internet; the monitor, keyboard and mouse are properly connected etc. We can almost think of the machinery behind the web today as if it were a newborn, having to be spoon fed and tended in a comfortable environment. Over time the web can be thought to mature to the Grihastha stage, where the tables are turned. Now it will be the responsibility of the machine based intelligences to provide a nurturing environment for humans to plug into. It is relatively easy to extrapolate the web gasket interface to support not just information transactions but also our physical needs (like medical monitoring). We will further discuss the evolving man-machine interdependence, and sustainable end-game scenarios, in Chapter 11.

When the web evolves to the state where it can readily accept Category 4 gaskets – we will have laid the foundation for the next step of Human evolution towards the Supra entity discussed in the previous Chapter. At the mental level, the web can now sustain us with all the memories and thinking processes we need – and the output of our work can remain in the web itself. Using the human cellular analogy, we now become similar to a neural cell that never needs to physically touch another neural cell to get signals – they get it through synaptic connections within a fluid medium (the new web). This new medium can reach out through thousands of miles and Exabytes (10^{18} bytes) of saved information about the human condition – and still remain instantly accessible from anywhere in this world.

Vast and anonymous as the web might be – we know from the Read/Write Asymmetry that it is not practical to just write things out to the web at large – especially information that is constantly being accessed by others. To match the granularity of the individual human, we will discuss in Chapter 9 the creation of the Machine Supra Cell (MSC) – a machine based intelligence that is close to our human intellectual capability, and can serve

as the sanitizing and buffering organism for everything that we would write back into the greater electronic medium. Just like the personal computer (PC) that allows us to interface to the internet of today – the MSCs will have the capability to carry on a vastly more meaningful interface to the rest of the web on our behalf.

Chapter 8
Brahmacharya for Humanity – a Pattern Search

"The whole visible universe is but a storehouse of images and signs to which the imagination will give a relative place and value; it is a sort of pasture which the imagination must digest and transform".
- Charles Baudelaire, 1821-1867, French Poet

In this Chapter we will take a look at the patterns and images that surround our existence, and try to put our own existence in perspective against the cosmic backdrop. As we look at the trends from the various aspects of our existence, we can map them onto the overall context of the Mahashrama cycle. Once we have looked at a range of patterns, from the microscopic to the astronomical, we can take the first intrepid step of projecting where Human existence is headed in the near and distant future. In this Chapter we will call on the Brahmacharya spirit to discover the 'Reality' around us – and put our imagination on overdrive to figure out where we might be headed. As aptly stated in the above quote from Charles Baudelaire the great french poet, "the images and signs" we find in the "visible universe" … "it is a sort of pasture which the imagination must digest and transform".

Mirror Mirror on the Wall

A useful measure of intelligence amongst humans and primates is a common household object – the mirror. At an age of about 18 months, human children are typically able to look at a mirror and recognize that it is themselves that they see. They then proceed to explore, with the help of the mirror, those parts of themselves that they cannot view directly, for example their teeth and the inside of their mouth.

Only a few species of Great Apes (e.g. Chimpanzees, Orangutans) consistently display such self awareness[1]. Other Apes, (e.g. Gorillas) seldom show this capability. Monkeys and most other animals (the jury seems to be still out on bottle nosed

dolphins) do not display this characteristic. Whether or not we accept this ability as a defining measure of sentience, it definitely points out that humans and a select few other primates are different in our perception of what lies behind the mirror. We have the capability to visualize and discern not only our surroundings – but also ourselves as it relates to our surroundings.

The image that we see is delayed, even if for a tiny split second, to allow for the propagation and perception times mainly within our eyes and brain. To take it a step further - when we visit a museum and view a trick mirror that makes us look tall or fat – there is still no doubt that it is ourselves that we see. As we look at all the imagery in the visible universe, we will need to work through many levels of time-delay, distortion and scaling to see the image that is truly relevant to us.

Patterns and Images around the Mahashrama Cycle

Looking at Figure 8.1 we see several key images and trends arrayed around the Mahashrama cycle. With our perception of what's behind the mirror, we have already begun the discussion on Individuality and Self Awareness (resident in the Brahmacharya segment). Let us work our way clockwise through all eight highlighted areas – in each of which we can discern one or more patterns or trends.

Individuality and Self Awareness

Using the mirror test, we have found that it gives us a measure of our own self awareness as it relates to our environment. With modern technology, we now no longer have to rely directly on our senses to tell us everything about the world around us. Today we have pictures of the Earth from space, and during the process of writing this book, pictures came in from Saturn's moon Titan (Robotic Explorers - Huygens, Cassini). Though coming from so far away, these pictures look vaguely familiar – as if they were taken on our own planet! Thus, as we work to sharpen our innate ability to identify ourselves within our environment, let us remember that our environment is not just our locality, or even

Earth – we are now emerging citizens of the Solar System, the Milky Way Galaxy and the Universe at large. The complexity and organization of a human – looking inward – is often referred to as a Microcosm. As we go through this cycle, we will see several patterns that extend all the way through our Micro-cosmic being to the Macrocosm of the universe.

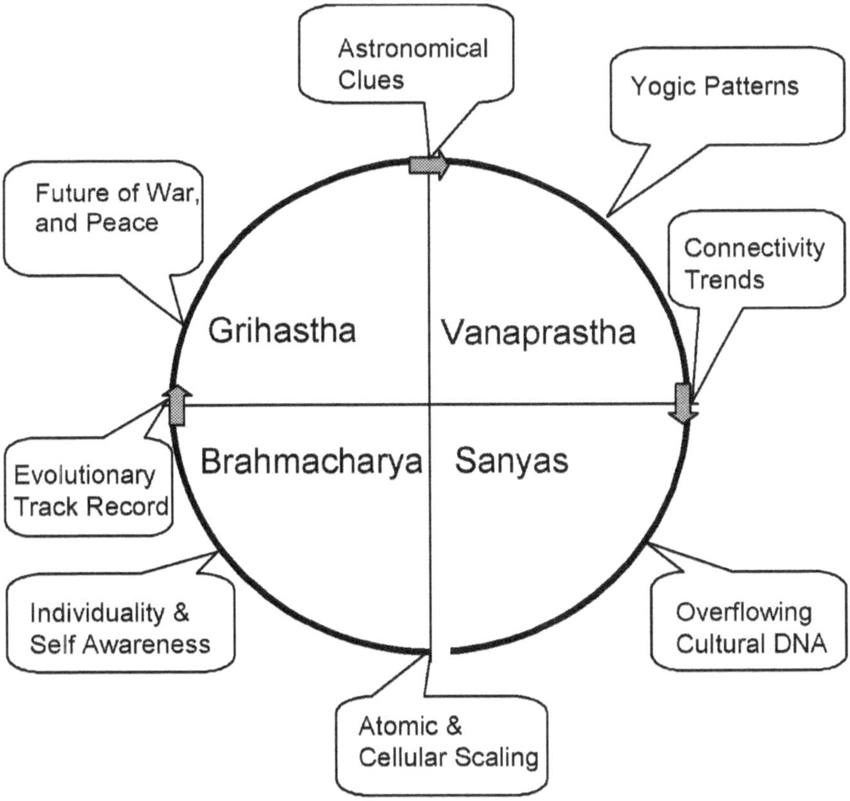

Astronomical Clues

Yogic Patterns

Future of War, and Peace

Connectivity Trends

Grihastha

Vanaprastha

Evolutionary Track Record

Brahmacharya

Sanyas

Individuality & Self Awareness

Overflowing Cultural DNA

Atomic & Cellular Scaling

Figure 8.1: Images and Trends around the Mahashrama Cycle

Evolutionary track record – for DNA, Species and Civilizations

Slowly but surely, over billions of years, the random process of evolution has resulted in a huge diversity of life on earth, including our own. By its very nature, for every positive change (typically a

mutation at the DNA level) that has survived and prospered, there have been hundreds that have not yielded a positive change - and subsequently caused the organism to be less fit and die away.

> *"There are millions of different species of animals and plants on earth -- possibly as many as 40 million. But somewhere between 5 and 50 billion species have existed at one time or another".*
> *- David Raup, in his book -*
> *Extinction: Bad Genes or Bad Luck?*

Thus, it is estimated that 0.1% or less of all species that ever existed are still alive today – and a great many of them are facing extinction. Humans, in our current form (Homo Sapiens Sapiens), have only existed on earth for about 130,000 years – a blink of the eye in cosmic terms.

We find ourselves again at an evolutionary crossroads. There are some key trends that indicate that the functionally advanced segment of Humanity will no longer be tied to our current biological existence at the turn of the next millennium. Let us take a look:

1. The human biological genome has been recorded, and large sections of it are getting decoded to decipher how they function. Designer genes are not that far away[2] Any change beyond the very superficial – could lead to a genetic sequence that is no longer biologically compatible with the rest of humanity. In evolutionary terms, the size of the Human brain appears to have been limited by the dimensions the mother's pelvis. With more and more births coming via C-Section, even this constraint begins to go away. Morals and ethics apart, the creation of Super-babies cannot be too far behind.

2. The human brain is exceptional in the animal kingdom. The human senses are quite average. Human muscular performance and speed are bested by a wide range of animals – including the household dog.

We are already learning to build artificial organs (e.g. heart-lung machine, dialysis machines, and artificial retinas). The day is not far away when humans can get a significant competitive boost by adopting artificial organs and muscles that retain a high level of performance over time, and can be replaced if needed.

3. We are already implanting tracking devices and pacemakers into humans. Over time it will become possible to implant electronics that could give us miraculous powers – like photographic memory or even direct electrical interface to computers.

4. Man's physical construction is not suited for extended space travel. Some combination of #2 and #3 above could put us in a position to morph into creatures that do not need an oxygen rich atmosphere and bulky space suits to move around in space. As it is, the earth only receives about a billionth of the total radiated energy from the Sun. Over time, we will outgrow our renewable energy budget here on earth just as surely as we will outgrow our reserve of fossil fuels. Space based engineering would become a necessity – and a space hardened subspecies of Humans seems very likely.

5. If we Humans fail to make the jump to space, our machine based creations (also known as robots) surely will. As their intelligence grows exponentially, over the next 20-40 years their intelligence could outstrip ours. It is one thing for man and machine to live together in close proximity in Yogic union – it is quite another for machines to feel that they are serving their colonial masters from across the voids of space.

Since we do not have billions of years for evolution to work its slow progress, it is logical to think that the major changes will need to be wrought by design. Otherwise, the fate of the current strain of humans could be fairly short lived – or worse we might be relegated to living the role of a curiosity in a machine driven

civilization. Not making the leap into space might mean that we get sequestered in this lonely planet called Earth – while our own biological and/or machine progeny enforce a "No Fly Zone" around us.

For most Civilizations in human history – the evolutionary track record for continued existence is indeed quite dismal. In cases like the Inca Civilization – most of their wisdom and technology passed away when they were conquered. With the headlong rush into globalization, a lot of regional languages and cultures are seeing their active practitioners minimized and marginalized. Even our current technology based civilization is currently at a crossroads. It is worth noting that the general learning from History seems to be quite clear – "adapt and transform, or die!"

Future of War

Beginning with the German V1 and V2 missiles during World War II – a clear trend has started in Warfare that is worth noting – and projecting into the future. The V1 Buzz Bomb was the first cruise missile and the V2 the first Ballistic missile. The innovations that started with these missiles were the first electronic guidance and flight control systems. Another event of note, coming at about the same time, was the outfitting of the B-17 flying fortress bombers with remote flying capabilities – spawning the category of military drone aircraft that could be used to deliver significant firepower.

The fortunate thing about the flight and guidance systems, as well as the remote control/communications systems on the drone is – they are subject to Moore's law. The flying fortress, with a gross weight of about 65,000lb carried less sophisticated electronics than the late '90s Predator Drone grossing at about 2,000lb. If we are able to extrapolate using Moore's law (which is slower in Military applications) – theoretically the same amount of electronic payload can be carried in a 2lb craft (about 1 Kilogram) in the 2020's and in the footprint of a hornet or bee in 20-30 years after that. These timeframes are admittedly approximations, but the mechanics of the battlefield are definitely changing. When dealing with human enemies – it will no longer be necessary to destroy

them as specific steps can be taken to incapacitate them. The loser of a battle may feel like this is a new form of individualized slavery. However, the conditions can be relaxed once the military objectives have been secured and the enemy disarmed.

Another implication here is – no place on earth would really be safe from some kind of infiltration. It is a sad fact of modern technology that defenses are much harder to fabricate than offensive capabilities. When offensive systems begin to verge on the microscopic, an immune system with accurate 'friend or foe' identification becomes critical to our functioning as individuals and society. If this is beginning to describe what the Human body does already – this will be a recurring theme while we continue examining other trends around the Mahashrama cycle.

Future of Peace

The future of Peace will indeed be about 'Eternal Vigilance' – especially when it comes to keeping up with the technology curve. When probes verging on microscopic can carry a lethal payload, it becomes important to know what belongs and what does not belong in our immediate vicinity. In public spaces, it will be up to the governmental authorities to make sure that safety and security is maintained. In times of strife, the task of maintaining the safety of a given population becomes so complex that the entire outdoor environment could begin to look like a closely meshed sensory network – tracking and correlating everything in our environment. This does put a significant strain on our Vanaprastha Network - which ultimately takes away from humanity's ability to work on those long term goals that can ensure success and prosperity for all of us.

The alternative, of course, is to create a zone of trusted entities; where anyone that has not yet earned the trust of the community is very closely monitored. The first thing that the arrested person will lose is their privacy – and that may be all that they need to lose! With the ability to monitor every single action of a person in public and in private – the convicted felon would carry a virtual 'wall' around them that could remotely restrain them from any harmful activity.

Astronomical Clues

As we look around us, it is hard to escape the fact that we are intimately dependent on the Sun as the primary source of our energy, and beyond that, to support the kind of environment here on earth that has given rise to our species. The Sun and the Moon have been the subjects of endless speculation on the part of Humans – as well as veneration and worship.

The first astronomical 'coincidences' that is readily apparent to casual observation is that the sun and the moon cover about the same arc in the sky. This is just about half a degree of arc. This registration is so close that depending on the relative location of the earth and moon in their slightly elliptical orbits – we can get a total eclipse of the sun (by the moon) or an annular eclipse! By itself this is not a great coincidence – but it illustrates that we live in a scalable universe, and that it is equally important to be able to scale both backwards (that which we revolve around) and forward (that which revolves around us).

To establish any level of periodicity, it is useful to have at least three data points. Let us consider the following facts about the sun and earth:

Earth Diameter: 12,756.2 Km (Kilometers, at equator)
>Ratio: 109.2
Sun Diameter: ~1,392,700 Km
>Ratio: 107.7
Earth Distance from Sun: ~ 150,000,000 Km

Table 8.1: Earth, Sun Diameter and distance ratios

The ratio in either case is very close to 108 – a number that was considered a symbol of divinity in Ancient India. Interestingly enough, the same ratio is roughly true between the diameter of the Moon and its distance from the earth. For those numerically inclined 108 is also the product of the first three positive integers to the power of themselves – i.e. $1^1 \times 2^2 \times 3^3 = 108$. These numbers would mean nothing, of course, if earth was too hot or too cold to

support life – but the fact that these ratios could be maintained and still give us the life sustaining environment that we have today – is quite remarkable. Mathematically 108 is also equal to $0^0 x 1^1 x 2^2 x 3^3$ - with the '0' term adding a whole dimension of meaning that we will develop in Chapter 10.

Scaling up (108x) from Earth-Sun distance, the distance we get is 16.2 billion Km or somewhat larger than the orbits of Quaoar and Eris (formerly known as Xena) – two of the farthest known planetary objects to orbit the sun. This number is probably a good estimate for the planetary part of the Solar system! Extrapolating down one step, the Earth's diameter divided by 108 gives us 118Km – close to what is considered the edge of space from Sea level. Is there any other significance to this quantity – perhaps in human or biological terms? The answer is a resounding yes, as we will see in a Chapter 10. There we develop a set of conclusions from these clues – which will connect our human existence from the microscopic all the way to the astronomical.

Yogic Patterns

Human biology loves the number two. Consider:

> Number of the same type of chromosomes in a cell: Two
> Construction of DNA: Double Helix
> Cell division ratio: 1:2
> Human body symmetry: Bilateral
> Human brain construct: Two interconnected halves
> Vision: Two eyes
> Human Locomotion: Two legs - and the list goes on …

Even outside of the human body – the number two seems to rule. Marriage is the union of two people. In most religions the relationship is between us and our God (Two entities). Games are usually played between two teams. Most legislative bodies are bi-cameral – like the US House of Representatives and Senate. The base for the numbering system for computing is also binary – or two. Apart from being of acute numerological interest – is there a significance of the number Two – or of the 1:1 relationship that is implied within it?

Let us take Human Vision as an example. We perceive a single stereoscopic vision using two sets of visual inputs from the two eyes. That each of us has a dominant eye can be proved with a simple experiment. Let us take a piece of paper and cut a hole in the middle of it (~ 1 Cm diameter) and hold it at arm's length. We then spy an object through the hole – and then slowly draw the paper closer to our face. Almost universally, we end up with the hole sitting right in front of our dominant eye.

From personal experience, the author's dominant eye happens to be the left eye. After a set of corrective (laser) surgeries the right eye happens to have better eyesight– but the left eye remains the dominant one. At least in this case, the synthesis of the author's vision seems to happen just fine – which leads one to believe that dominance is more a matter of habit than functional competence. Too many human relationships seem to end up with one person trying to dominate the other. With adequate Yogic training the dominance could come and go, but the functional capability to synthesize the totality of the situation should not be effected. A similar dominance pattern can be seen between the two sets of genes inside our chromosomes, as shown by the pioneering experiments of Gregor Mendel (1822-1884).

If the ability to combine two dissimilar components into a combination that is united in its identity and function is so common in biology – why is it that humans have such a difficult time with the concept? Human behavior too often looks like individual eukaryotic cells before they came to develop their own Yogic skill. Using this skill, cells regularly transmit signals between themselves which determines which set of genetic features to turn on, when to undertake cell division – and at the appropriate time, when to die!. This kind of Yogic connection is hard to come by amongst humans today – at least until we have in place the trusted (and technologically sophisticated) Vanaprastha Network that we discussed in Chapters 4 and 5. The Brahmacharya and Grihastha will need to do their bit as well – opening up a part of themselves to the trusted vanaprasthi (or Guru – in Indian spiritual terminology) for consultation and guidance.

An amazing metamorphosis happens when eukaryotic (nucleated) cells, through the process of binary cell division, split into hundreds of different cell types and then self assemble into a complete human being. They do this with just the genetic information contained within their chromosomes. The somewhat equivalent process of self organization in humans – depends on the Vision of a successful future that the Sanyasi has inculcated into the Vanaprastha Network. We all know the power that John F. Kennedy's vision of 'Man on the Moon' had on getting the US space program energized to reach that lofty goal. The Moore's law that we have used as a reference rate for the advance of miniaturization in electronics – is also a great example of a self-fulfilling vision.

Even the Mahashrama Cycle can be thought of as two successive binary divisions. The first is the vertical line that divides along the center. This constitutes the first existential separation – the one between what exists (i.e. our physical presence), vs. why it exists (i.e. our Purpose). The next is the horizontal line that divides the potential (ideas, skills, vision) from the actual (instantiations with a high economic footprint). The glue that joins these diverse halves together is the power of Yogic Union – where two dissimilar halves can come together and perform as one.

The enemy of Yogic Unity is the existence of barriers that are put up to isolate ourselves from the endless deluge of events and interactions that life throws at us. The barriers are a necessary evil – because unless one really isolates themselves (say a newly married couple on their honeymoon), the typical human will not have the bandwidth to support a thoughtful constructive dialog with every other person we come in contact with. Here too, we have exceptions of course – especially with great spiritual leaders (like Mahatma Gandhi) who are able to reach across vast multitudes with the power of their yogic spirituality and purpose. In the presence of barriers all around us – it is all too easy to define ourselves in terms of the boundaries we see around us – whether in the form of class, creed or religion. For yogic connectivity, however, what is more important is determining which of these barriers around us might lead to the presence of a kindred soul at the other side with whom we can build a trusting synergistic relationship.

Connectivity Trends

Unfortunately, there is nothing like a great natural (or manmade) disaster to focus our attention on the heroics at the individual level and the deficiencies at our collective level. In late 2004 we were still counting the losses from the devastating South Asian Tsunami. Certain patterns emerged, which we will use to illustrate the connectivity trends amongst humans.

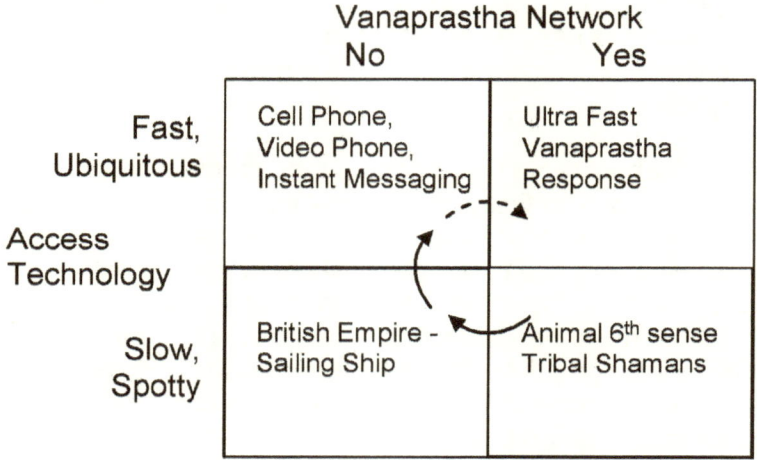

Figure 8.2: Human Connectivity trends

The first set of stories we hear are of animals and primitive tribes that were somehow able to detect the patterns that indicated that something was horribly wrong – and were able to move out of the disaster area in time to avoid the worst of the calamity. One of the more credible stories is of Victor Desosa, the village headman of Galbokka in hard hit Sri Lanka[3]. During his active career, Victor had been a merchant marine sailor, and had experienced first hand the earthquake in 1982, when his ship was outside the harbor of Valparaiso, in Chile. When the characteristics of the ocean on the day of the Tsunami reminded Victor of his Chile experience, he sounded the alarm and is credited with potentially saving thousands of lives. From the same catastrophe, we also have stories of prescient animal behaviors that seems to have saved them (and at least in one case a human child) from the worst of the waves.

If we look at the 2x2 chart in Figure 8.2; Victor's heroics would fall in the bottom right box. Galbokka did have a Vanaprastha Network (Victor was quickly able to mobilize his village) – but there was no use of modern technology for communication. As humans become more individual and independent – one of the first things we tend to lose is our tribal identity. In doing so, we tend to rely more on our independent rational ability to make sense of the world around us. The British controlled their very substantial Empire through the use of written proclamations (on paper) that were then dispatched around the world on sailing ships. The local Governors were given a lot of autonomy – but there was never any illusion that the far flung empire was being run for the common good of the population being governed. Before the advent of wireless communications, the British Empire's communication systems would fall in the bottom left corner of Figure 8.2.

With the widespread adoption of wireless technology at the turn of the millennium – cell phone coverage has been extended to cover the world's population centers and much of the rural areas as well. Initially there was just voice based telephony, and then came SMS (Short Messaging Service) – and now we also have picture and video capabilities on the wireless potable device. With '911' emergency dialing capability we have just the hint of a Vanaprastha network – but it works only one way. If a Tsunami is coming your way, there is not yet a system that allows the cell phone to be used as an Early Warning System. If we get into an accident, our typical cell phone (circa 2006) will not know enough to call for help. If there is a poisonous gas leak (like the one in Bhopal, India in 1984) there is no way to chart the path of the cloud of gas and warn those at risk, and alert them to possible countermeasures. Hence, our current situation (2006) puts us in the top left corner.

Once we have in place a responsive Vanaprastha Network we could now move to the top right corner. This same cellular access technology now becomes much more of a friend, especially if we go by the Proverb: "A friend in need is a friend indeed". To do this effectively, the personal communication device (next generation cell phone) will need to be aware of things in our immediate vicinity – e.g. the state of our car (is it in a crash?), or

even the state of our health (are we having a heart attack?). We can envision a scenario where a driver in a car or the pilot in an airliner is incapacitated, and the vehicle automatically goes into safe mode and prevents the vehicle from running off the cliff (or into tall populated buildings!). The Vanaprastha Network is all about service, and about people who 'care' for the population they serve. With a population in most countries that is getting older and older – we could see quite a significant medical services industry springing up to provide 'Just in time' intervention and support.

The connectivity trend works both ways – just as the person in need of assistance can rely on their connectivity to get help, we must also get the situation specifics to the vanaprasthi who can provide the help. One of the poignant symbols of the Cold War, the Red Telephone, was supposed to be available for instant and direct access between the President of the US and his counterpart in the USSR. Thanks to today's ubiquitous cellular technology – every vanaprasthi can be so equipped – thus enabling them to respond to a serious problem in a timely and coordinated manner.

Overflowing Cultural DNA

If we look at the amount of information coursing around in the Web – there is no way that a single human can even begin to look at all of it – leave alone digest and assimilate it all. The fact of the matter is, the human brain's intrinsic capability has not developed very much (if at all) in the past few thousand years, and yet the torrent of information available keeps increasing at an exponential rate!

It is estimated that at the end of 2002, the entire web content (including the places not frequently visited) held about 91 PetaBytes of Information For reference a PetaByte is 10^{15} Bytes – more information than the all the Synapses in the human brain can theoretically hold. If no single human can even look at this amount of information – what are the chances that it can all be digested and simplified down to a single Cultural Genome for Humanity?

Humans have adapted to this information overload by creating narrower and narrower fields of specialization. As a result Human knowledge is growing fast – but not necessary as a contiguous and meaningful whole. To each of us the sea of information looks like a planet girdling ocean and we get to see and influence only a very small part of it! Occasionally, we will send out a probe in the form of a web query and get back a few thousand approximate answers to work through! Admittedly, the search engines have improved tremendously over the past decade – and typically give us a very good selection to work from. But finding the pertinent answer to a complex question is still a very time intensive effort.

The same pattern of information overload is repeating elsewhere – to the point that all of human existence appears to have become a huge information overload. The Sunday papers in a mid sized Metro area (Phoenix) now seem to have as much heft as a phone book! Hundreds of channels of TV are available over Cable and Satellite – and more are on the way. A lot of our daily activity seems to be just rushing from one place to the other – doing the same repetitive activities over and over again.

Going back to the Mahashrama cycle, the information overload and a fast beat rate are typical of the Brahmacharya stage of our existence. So are a high degree of competitiveness, and the corresponding feeling that we are running ragged just to keep up. Coming back to the discussion on the Web, the needs that it could potentially satisfy over time are summarized in Fig 8.3. As usual, we start with the Wide Array of Information available in the Brahmacharya segment.

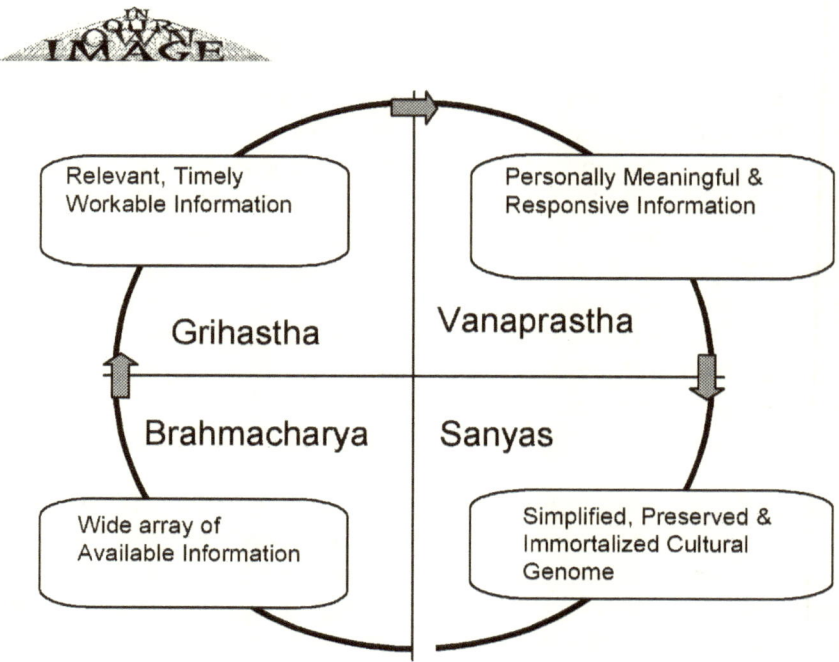

Figure 8.3: Projected evolution of the Web

With the improvement of the search engines, the relevance of the information that we get back is increasing, but there remains a huge scope for improvement. If we are working on a project, and seek information to work into it (say in the writing of this book); it is useful to know what the boundaries of the current information are – so that we can build at the periphery. The amount of context required to define what specific information we seek is typically far higher than today's word based query – and requires that the search parameters include what information we already know. For example, let us say that we'd like to find a Quote or a Proverb that carries the core idea of every Chapter in this publication. How would we go about designing a query that gave us the top 20 selections? Yes, there is an element of our individual endeavor that now needs to be passed on as a parameter – and the search would need to be in the 'Meaning' space covering multiple languages.

When the web moves to the Vanaprastha stage, the information available has all been connected by meaning – and is beginning to look like a contiguous whole. Yes, there are many repetitions, anomalies and errors in the content – but it appears to hang

together. Now, when we do a query – the parameters passed include not only the workspace we are looking to build – but also our individual context so that the response can be tuned to our own specific situation. This requires some appreciation of our specific situation – and gets back to the 'Connectivity' thread that we were discussing in the last section. Let us use an example to illustrate the situation.

A family of four drives into an unfamiliar town and is thinking about a place to eat. The father is a diabetic and needs to find something high in proteins and low in carbohydrates. The mother is an aerobic fitness nut, and having missed her workouts during the day wants the maximum bang for the few calories of nourishment that her web based 'conditioning coach' would allow her. One of the kids has a hankering for pizza – and the other would like pasta. The web response assimilates all this information, checks against the menus of nearby restaurants – and suggests the best fit (with one backup option). When the family arrives there, the order has already been placed. They have a quick meal; and the 'conditioning coach' even remembers to tell the mom to avoid the bacon bits at the salad bar. It might take a bit of getting used to – but most individuals would likely not object to having their needs anticipated and serviced in a customized and thoughtful manner.

If such a responsive web environment reminds us of the attention we used to get from our mothers when we were little children – here is some food for thought. To an infant, the mother is quite close to being the entire universe that he or she interfaces with. Would we allow the web to envelope us in such a fashion? Or will there be avenues of personal endeavor that will continue to be challenging and unique to us – in spite of the best help that the web may be able to give? On the opposite camp, will there be people who are content to just feed off of the system and not put in any effort to advance the greater cause of humanity? And how would the economics work if we were to implement such a system and roll it out to an entire population?

The last stage of the web, we have talked about already, is where the totality of the human experience is collected, made internally consistent (allowing maybe two threads for some topics

for diversity of viewpoint), preserved and immortalized. We have also indicated that a very realistic way to bring out the combined experiences for posterity is to make sure the Human Cultural Genome can be re-enacted and re-lived.

Fundamental Periodicity: Atoms → Cells → Humans → Supra Entity?

We noted in Chapter 6 (Figure 6.9) how the Human Engineering footprint is scaling over time – and how it relates to the full range of the Universe. At this time, let us zero in on the mid range of this chart to take a look at how the Human body itself is constructed. At the low end let us take the Fermi length (fm) which equals 10^{-15}m, the typical size of atomic nuclei. Figure 8.4 is also a logarithmic chart in the vertical (length) dimension. However, the bars are spaced further apart for more precise positioning of the relative sizes.

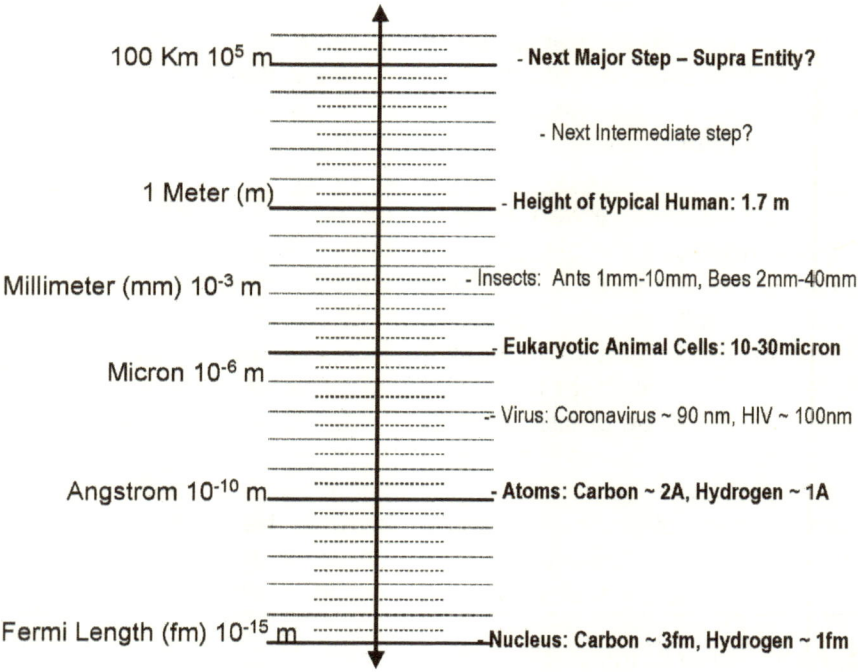

Figure 8.4: Relative Scaling: From Atoms to Humans and beyond?

The first thing that we note here is a major periodicity – all the way from the sizes of the atomic nucleus, to the atomic sizes, the cells that we are made of, and finally our own dimensions. In each case, the periodicity is almost exactly 100,000x or a factor of 10^5. Interestingly enough, the name for 100,000 in Sanskrit is 'Laksham' which also means 'target'. The ancient Indian scholars would not have known about the 'target' hidden in the ratio of an atomic nucleus to its atomic size. Seeing that the same 'target' repeats itself as we move from atoms to eukaryotic cells, and then on to human beings; one must wonder if our ancient forefathers had insights into the nature of the universe that we are just begging to uncover using the tools of science.

If we push this forward another 100,000 we have an entity in the order of 100Km – could there be something magical at about this dimension? The observant reader will probably remember that we came across just such a number by scaling the astronomical factor (108) downwards. The astronomical scaling suggests that there is a next step at about 118 Km that fits the periodicity of the heavens. Now we are finding a number in the same range that fits the periodicity of human construction. For now, let us keep a placeholder at that dimension for the Supra-human entity – a hypothetical being of a scale where humans (or equivalent machine intelligence) would be as numerous as cells are in a human body (i.e. ~10^{14}). For now, let us hold the thought that each meta-step of our existence as we move towards a Universal Scale Intelligence could be separated by the 'target' dimensional scaling factor of 100,000. As noted earlier, we will dig farther into the potential messages behind both these periodicities in Chapter 10.

If we look closely at the pattern in Figure 8.4, we see another smaller periodicity halfway between the 10^5x increases between each major step. Almost directly between the atomic scale and the cellular scale we see intermediate life forms like the virus. Also, almost smack in the middle between eukaryotic cells and humans we find the numerically most successful and abundant of all multi-cellular creatures – the insects. This intermediate level of periodicity does suggest that there may be an in-between stage of encapsulation for humans at about the 300m size range – which we refer to as the sub-Supra.

Uniformity to Universality

If we take the pattern form Fig 8.4 and extrapolate it out till we reach the limits of the Universe, we can see the kind of headroom that we have for future growth. Beyond the Supra level, we will define four more levels – Super Supra, S^3, S^4 and S^5 at the same degree of periodicity (10^5) as we have observed from Atoms to Cells; and from Cells to Humans.

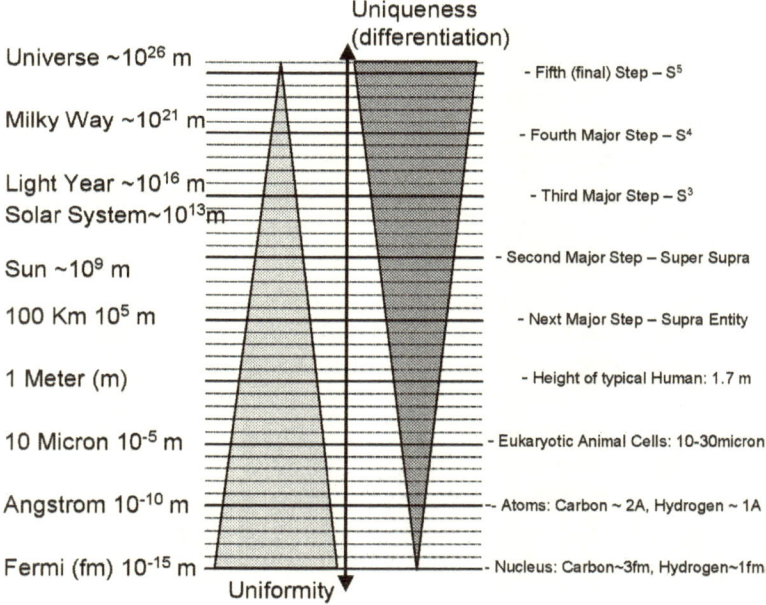

Figure 8.5: Atomic and Cellular Periodicity extended to the Universe

At the atomic level, there is very little to distinguish one atom from another – and they can be said to have a very high degree of 'uniformity' from one unit to the next. At the cellular level there is still a lot of uniformity, but we begin to see some differentiation or 'uniqueness'. We know that several hundred cell types can be

expressed from the same basic genomic content inside each one of our human cells. Even from cell to cell (e.g. Neurons) there can be significant differentiation in size and complexity.

In Figure 8.5, Uniformity is represented on the left as a triangle, broadest at the bottom (Atomic level) and tapering off at the top end. On the right, we have and inverted triangle representing uniqueness - that is widest at the level of the Universe, and tapers down to almost zero at the sub-atomic scale. In the middle, we have some degree of both – i.e. the units are interchangeable to a degree as well as possess a certain amount of individuality and uniqueness.

At the human scale, we have significant amounts of both Uniformity and Uniqueness – but in the big scheme of things we are still more similar than different. Most human genetic material differs from each other by significantly less than 1%, and we are all more or less of a similar physical size and have similar intelligence attributes. If we extrapolate to the scale of the Supra – it would appear that there will be an approximately equal degree of Uniformity and Uniqueness. This is something to keep in mind as we work on laying the foundations for our next level of existence in Chapter 9 – In Our Own Image. (For those with a numerical inclination - in binary 1001 = 9)

Chapter 9
The New Grihastha - In Our Own Image

Tiger, tiger, burning bright
In the forests of the night,
What immortal hand or eye
Dare frame thy fearful symmetry?
- William Blake

In this chapter we will look at what it takes to build towards a higher level of Human Civilization that is consistent with the scale of the Universe that we find ourselves in. We have discussed several distinctly higher stages of organization for intelligences (e.g. Supra, Super Supra, etc.) that we can aspire towards. A first step towards building up to these lofty entities is the creation of the Machine Supra Cells (MSCs). This chapter will focus on defining the key behavioral characteristics of our machine progeny, as well as the enhancements we need to human behavior patterns so that we can together build a highly scalable social fabric.

We have noted several trends from the last chapter – which we will incorporate into a high level plan. Technology is sure to take us on a wild and sometimes scary ride. The goal here is to minimize the threats and maximize our opportunities for success – while improving the confidence that we will actually survive this thriller! William Blake in the quote above talks about the creation of the tiger – which brings to mind the type of challenges we face as we work towards the creation of our Supra Human future. The symmetry we would like to achieve here is a balanced execution through the engine of the Mahashrama Cycle, and the building of a productive yet secure future where our human and machine progeny can live together and prosper. The potency of that which we are setting out to create is indeed fearsome, and it is natural to feel a little nervous at the magnitude of the undertaking. Should we take the 'dare'? We will see later in this chapter that we really have little choice in this matter – and waiting too long to build up the first Machine Supra Cells (MSCs) can have dire consequences for humanity.

Moore's Law

One of the most respected visionary technologists as of this writing is Gordon Moore – co-founder and one-time Chairman and CEO of Intel Corp. In 1965, he made an observation that the number of transistors that engineers could find a way of fitting into a given area on an Integrated Circuit (IC) had doubled every year since the invention of the IC. He also predicted that the trend would continue for the foreseeable future. The doubling trend has slowed down a bit lately –rising first to every 18 months – which was incorporated in the revised version of Moore's Law. Engineers have compensated for this slowdown by finding ways to create bigger ICs on bigger wafers – which as of this writing measure a full 12 inches (300mm) on a side. For those of us not familiar with the IC manufacturing process, the wafer is a thin slice of semi-conducting material (typically Silicon) upon which multiple ICs are imprinted, and then later separated out and packaged.

A simple way to make sense of this complexity is to track the 'Generation' of the technology as measured by the length of a switching channel within the smallest transistor that can be fabricated. In 1985 (the year that the author joined Intel) the prevalent technology was at 1.5 microns, and the world was getting ready to welcome the 80386 processor from Intel – which was initially manufactured on a 1.5 micron process. As of 2005, the prevalent process technology node was 90 nanometers, with the expectation of a transition to 65 nanometer products over the next year or so. This represents about a 16x decrease in the linear dimension (2^4x) and a 256x decrease in area per transistor (2^8x). This data suggest that in the intervening 20 years the density has doubled eight times, or once every 2.5 years. To be consistent with recent history, let us take 2.5 years as a rate that we can use for future projections – at least until we get to atomic proportions.

In addition to process dimension, another feature that needs to be tracked is the maximum size single IC that could be engineered. For the sake of discussion – let us use the size of the wafer on which the ICs are built as the theoretical maximum size for an IC. In 1985, the prevalent wafer size was 4 inches. As of 2005 the wafer size had grown to about 12 inches (300mm). Practical

concerns typically limit us from building wafer scale IC's – most notably the number of defects that are tolerable in a given working circuit. Increasingly, some ICs (especially memories) are designed to be usable in spite of a certain number of defects. As the ability of designing defect resistant circuit rises – along with the ability to parcel out parts of a greater computation into parallel execution units – it is expected that more and more of the wafer will become usable as a single IC. Again, the theoretical limit here remains the size of the wafer – which, if it maintains its current rate of growth, could yield a platter more than 10 times the current diameter in 50 years. For practical purposes individual ICs have tended to remain below about 200 Sq mm, or about 14mm on a side – approximated by the dashed line just above the Centimeter line. Another practical constraint has been the size of the 'reticle,' or the maximum dimension that the optics that make up the photolithography process can expose at any one time.

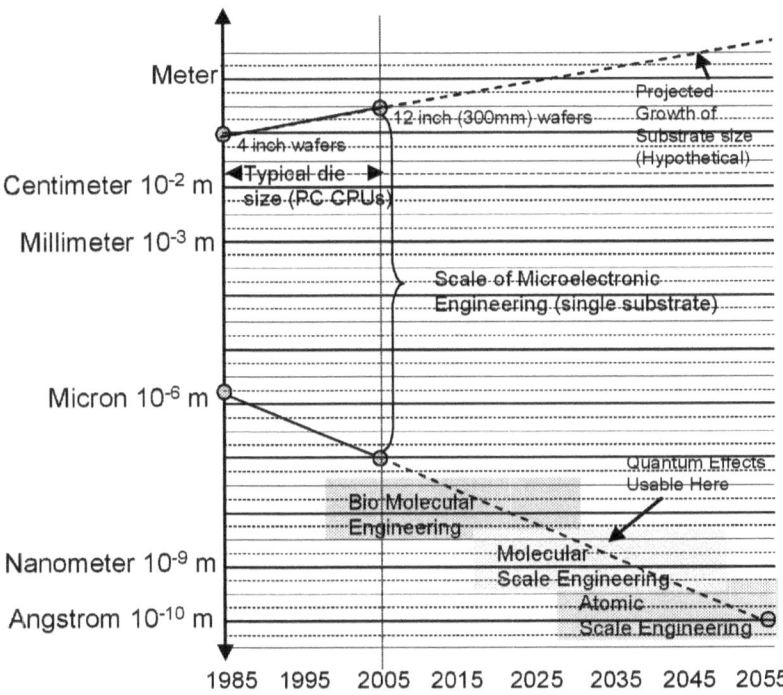

Figure 9.1: Technology Generations using a 2.5 year doubling rate for Moore's law

In Figure 9.1, we have each 10x range sub-divided into four to show greater resolution. Each small division is slightly less than a factor of two (~1.8x), with 2 divisions approximately 3.2x and 3 divisions ~5.6x. The 90nm process technology of today is close to the size of Viruses in nature – which are nothing but complex organic molecules. Hence we are almost to the dimension where bio-molecular engineering for computing purposes could become practical. In the lab, we can already build simple genomic sequences from scratch – without the need for having to take something that exists and altering it.

By the year 2025, at the current rate of progression, we should be within striking range of doing molecular scale engineering – using inorganic molecules (which tend to be smaller than organic). At about this scale, quantum effects become quite significant – and the technology could be used to make both conventional and quantum computers. By the end of this timeframe (i.e. 2045-2055) we should have the technology to engineer devices a single atom at a time!

Translating Linear Process Dimensions to Computing Capability

At the current point in time, we know the approximate capability (in MIPS or Million Instructions per Second) that a circa 2005 PC (Personal Computer) is capable of. In our conservative estimate, the processing power just keeps up with the complexity of the IC, which is inversely proportional to the square of the linear dimension. If we peg the performance of a typical PC today at about 3,000 MIPS – in 25 years this grows to 3,000,000 MIPS. In 13 years after that we exceed the commonly held approximation of the human brain -10^{14}MIPS [1,2]. This is a middle of the road estimate, with most estimates ranging from one order of magnitude lower – to two orders of magnitude higher. In the 38 years it takes us to get the personal computer to this level of complexity, our ability to engineer microscopic sizes would be at least down to the molecular scale – and about to transition over to the atomic level!

If we are willing to go for an expense level slightly higher, say that of a car, then the same amount of processing power can be acquired in just 25 years – roughly one generation in human terms. Having a house with the housekeeper built in might be a $25K option on the builder's list. A few years after that, having a car with a chauffer built in might be a $10K option. But - does having the human equivalent capability available mean that we have captured the essence of the human being in computer form?

Not quite! This is because intelligence is not a single scalable quantity – just as two youngsters scoring the same IQ in an Intelligence test are not interchangeable. There are aspects, especially requiring numerical computation – in which computers are already much better than humans. There are some others, like playing chess, where at the current state of development, computers are playing at the Grand Master level. There are still others, like spoken language comprehension, where a three year old child easily bests a computer today. There are yet other areas, like consciousness, where someone programming a computer would not even know where to begin!

One look at the Figure 9.2 and the Vanaprastha reaction might be – what a great example of Yogic Compatibility! A Grihastha reaction might be one of relief – since computers inherently are poor in the abstract human skills – this gives us a set of useful growth areas in life even if computers dominate everywhere else. To the neo Conquistador this might look like a new domain to claim – which properly tended could give them world domination. To the visionary Sanyasi – these are pieces of a puzzle that could be utilized to bring about the best possible supra-human civilization.

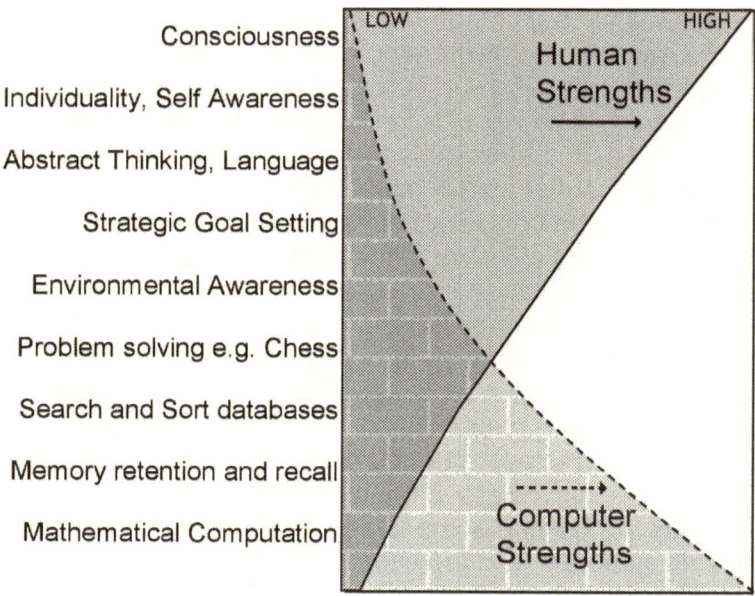

	LOW	HIGH
Consciousness		
Individuality, Self Awareness		Human Strengths
Abstract Thinking, Language		
Strategic Goal Setting		
Environmental Awareness		
Problem solving e.g. Chess		
Search and Sort databases		
Memory retention and recall		
Mathematical Computation		Computer Strengths

Fig 9.2: Differences between Human and Computer Intelligence

The first thing that the vanaprasthi will realize is that a sustained conflict between the two sides would seriously hamper any chance of a great and glorious future. Even conflicts between humans and machines working together on both sides are far more survivable than one that pits machines and humans against each other. The best solution, of course, is to build in an internal process for peaceful resolution of any conflict that does arise – lending criticality to the skills of Yogic Empathy in developing and educating our supra-cellular progeny.

The next thing that becomes apparent is that computers scale differently from humans. It is much easier for a thousand computers to break up and work part by part on a particularly chunky computing problem – especially since they are so similar in their internal execution. Humans don't scale so easily – which makes the Vanaprastha Network a particularly interesting challenge. Connected through the web - computing and storage are particularly scalable – a feature that is very useful as we chart a future that takes us (and our machine progeny) to an entirely new level.

<u>The Warp and the Weft – the Fabric that would Bind us:</u>

Borrowing from the art of fabric making – the warp and the weft are respectively the threads that run along the length and the breadth of the fabric being woven. It is the systematic binding of these individual threads that gives a piece of cloth its two dimensional feel and a strength hundreds of times higher than that of any individual thread. If the same set of threads were put in a disorganized pile on the floor they would neither have the durability nor the strength of the well crafted fabric. As we look into the future with our two sets of progeny – human and machine based – a curious pattern emerges as to the nature of these two types of threads.

In Figure 9.3 we take a look at how human complexity and attention span lays out relative to each other. The vertical axis is complexity, which we will measure in terms of the basic unit that constitutes us – the cell. At about 10^{14} cells, human beings are quite complex entities. On the horizontal axis, let us lay out the number of cycles of information processing that the human can perform. Human neurons can fire up to a maximum of about a 100 times a second – which means that an individual human working at maximum processing capability has available to them a total of 10^7 cycles in a given 24 hour period. Of course, the average human would not typically be able to apply anywhere close to that many internal cycles to a task at hand on a continuous basis – and our need for regular sleep puts an effective ceiling on how many cycles we can apply to a task at any one stretch.

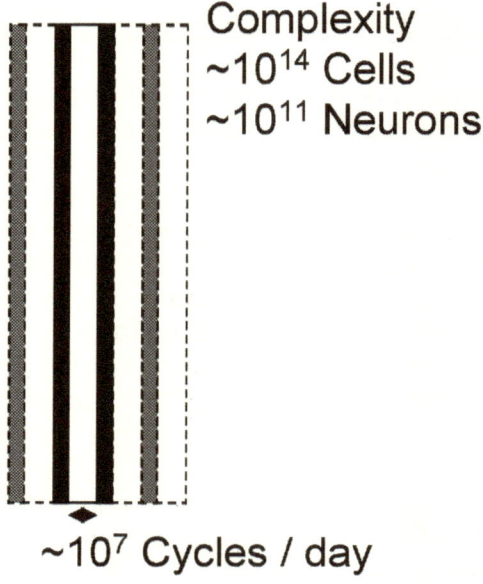

Complexity
~10^{14} Cells
~10^{11} Neurons

~10^7 Cycles / day

Figure 9.3: Task Complexity Cycles – Humans are tall and skinny

When we plot the complexity vs. the task specific cycles against each other, we find that humans are quite tall and skinny – i.e. that we can apply a lot of complexity to the task, but not for very many cycles. Even if we wake up the next day and apply ourselves to the task – the background processing that has happened during the night means that we are now a slightly different entity – another day older, another day wiser. If we apply the same kind of metric to the Machine Supra Cell the first stark contrast that we face is that the internal execution cycles now start getting counted in terms of Billions of cycles per second – or about 10^7 times faster than humans. We know that the circa 2005 Personal Computer (PC) can be pegged at about 3×10^9 in the 'instructions per second' measuring stick. Counting the number of transistors (mostly memory) in a typical PC, we come up with a number also in the same range. The way computers are designed today – especially for the mission critical tasks – it is possible to have a mean time between failures (MTBF) of years – i.e. the machine could go on performing its programming for years without needing a break. If we projected out to the day of the Machine Supra Cell (MSC) and

made a middle of the road (typical) estimate; i.e. that the higher processing capability comes mainly from complexity increase (1000x) but with some amount of speed increase thrown (33x) – the computing 'Attention span' begins to take shape.

With a 'Mean time between failures' (or MTBF) estimate of about 10 years we see in Figure 9.4 that the MSC form factor is really short and squat.

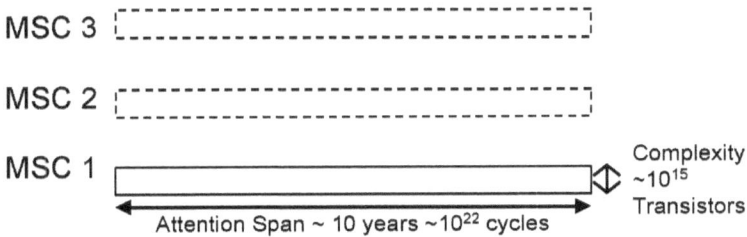

Figure 9.4: Task Complexity Cycles: MSCs are Short and Fat

The interesting thing is, both form factors are off of square by about the same amount – 10^7 – but in opposite directions! A well engineered fabric which ties the humans and MSCs together as the warp and weft of a future supra-human civilization has a much better chance of building itself up into a cosmic level presence to be reckoned with – than we can hope for with either constituent individually.

Three scenarios for Speed vs. Complexity increases in computing

The assumptions we have just described, with a relatively small increase in speed, and most of the 10^8 MIPS coming from additional complexity - is our typical scenario for developing the complexity level of an MSC. The conservative estimate, as described earlier in the chapter, is that from this point onwards, all of the processing improvements come from complexity increases (which track Moore's law); and speed increases, if any, will just go towards managing the added complexity.

There is an optimistic scenario as well – which is much less likely. The optimistic scenario says that the road to human equivalency will come equally from speed and complexity increases. If possible, this would get us the fastest to MSC scale processing power. Recent trends in CPU design suggest that for the reasons of power dissipation it is better to have two or more parallel processing units than a single super fast processing engine. It is anticipated that as performance increases, this problem of thermal over-heating will only get worse. Hence, for our discussion on the relative form factors of humans and MSCs the likely scenario we will use is that for every 100x improvement that comes from complexity increases, we will have only a 10x improvement coming from speed increases. To summarize, the time for PC class devices to develop human equivalent processing power (10^8 MIPS) in the three scenarios calculate out as follows (after 2005):

Conservative Scenario: 38 years, i.e. 2043 AD

Typical Scenario: 25 years, i.e. 2030 AD

Optimistic scenario: 19 years, i.e. 2024 AD

If complexity were the only metric by which we would mark the arrival of the MSC, the birth of the first MSC would likely be bounded by the range given above (2024AD-2043AD). In the author's assessment, the level of processing capability is just one of 18 different attributes that we will need to develop and integrate into a single functioning system for us to mark this defining moment in human accomplishment! This list is indicative of the kind of capability that the MSC will need to be successful in each segment of the Mahashrama cycle, and also to navigate between the segments as the opportunity arises – and is not intended to be exhaustive. A word of guidance for the intrepid reader who might feel dizzy as we work through the details here – if for some reason we lose you, let us regroup again at the end of this Chapter.

Cellular common denominator – Attributes of the MSC

As we build a future where humans and machines can live synergistically together in a supra cellular fabric – it becomes important to understand what the minimum qualities are that

define the attributes of a Machine Supra Cell (MSC). The obvious answer in this case is to consider a human being as a functional entity around which to model the machine intelligence required of the MSC. Figure 9.2 indicates that it will be very difficult to have an exact interchangeable mix of skills between a human and a machine – but we will continue to take 10^8MIPS as a rough equivalent for a definition of a MSC. Below that level, we can have aspects of a full supra-cellular functionality – but we are better off thinking of these sub-critical computing functions as subsets or stepping stones towards a full supra cell. In Figure 9.5 the four Mahashrama segments are augmented with a set of four 'elevators' that are instrumental in moving people from one segment to the next. In the list below the following abbreviations are used to indicate which segment or elevator each of the MSC attributes are drawn from:

V – Vanaprastha;
S – Sanyas

B – Brahmacharya
G – Grihastha

YU – Yogic Universality
TVI – Transcendental Vision, Identity
SS – Self Sufficiency
LCC – Love, Caring, Community

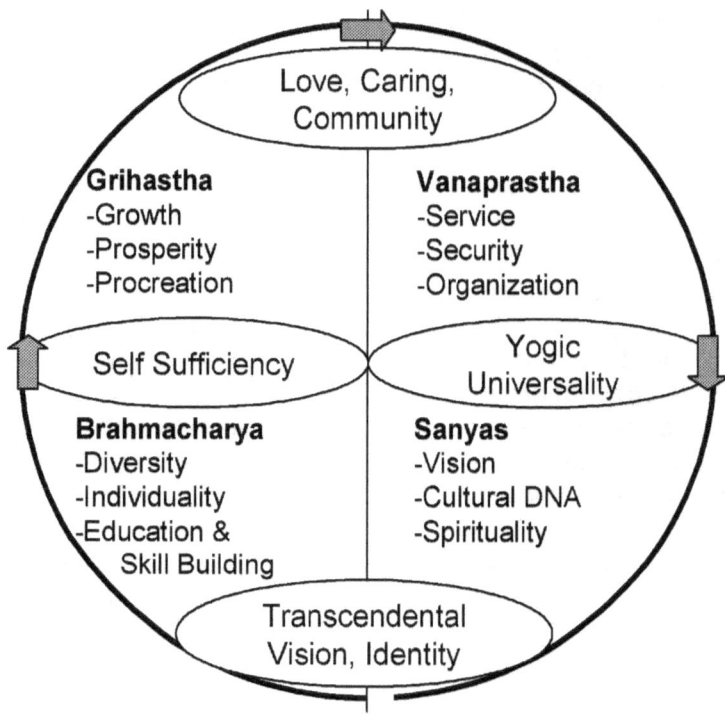

Figure 9.5: Mapping the Supra Cell design attributes onto the Mahashrama cycle

Our list starts with the Vanaprastha segment attributes, and works its way clockwise through the Mahashrama Cycle. The 18 MSC attributes are as follows:

1. Supports categories 1 – 4 gaskets for Yogic connections (V)
   a. Transparency of operation under gasket interface
2. Registration and certification of accomplishments (life registry) (V)
3. Ability to have higher degrees of freedom when it is fully connected (YU)
4. Ability to combine with another supracell and operate in yogic union (YU)

5. Ability to suspend all operations and go into 'Saved' state (S)
   a. Encapsulate intermediate learnings and disseminate to rest of Supra being
6. Immortality (S)
7. A genetic identity – typically the human cultural genome (TVI)
8. Personal area of interest and development – may or may not be connected to their Citizenship (TVI)
9. Ability to develop new capabilities and learn from its mistakes (trainable) (B)
   a. Takes new learning and is able to do a contextual merge back into original database.
10. Ability to resume operations from a saved or suspended state (B)
11. Ability to keep performing at some fail-safe level even if totally disconnected (SS)
12. Ability to monitor own environment for safety and productivity (SS)
13. Well defined spheres of influence and action (G)
14. Localized or distributed processing capability of ~ 10^8MIPS (G)
15. Ability to reproduce self to another compatible hardware base (G)
16. Citizenship in a collective – e.g. company/individual they work for; and a collective identity – typically values, purpose, vision of the collective (LCC)
17. Ability to scale operations across cell borders (LCC)
18. Ability to recognize and act on cell to cell transmissions as to what attributes to express and what functions to do over time. (LCC)

Vanaprastha Attributes for designing an MSC

1. Supports categories type 1 – 4 gaskets for Yogic connections
   a. Transparency of operation under gasket interface

We have already seen (In Chapter 7) that it will be possible to create different categories of Gaskets (categories 1-4) that allow a human to connect seamlessly into a machine environment. Category 1, we have today – defined as the direct physical interface (joystick, steering wheel) that we interface directly with our muscles and senses. Category 2 – we are beginning to see more of, is the ability to just think-it to achieve an action - and bypass the physical actuators. Category 3 – is the ability to have our memories also resident on the machine. Finally at Category 4, all of our thinking processes are also machine resident – and we interface only under the power of our consciousness.

Working cooperatively through a set of problems and issues – with a Master Yogi (Senior vanaprasthi or sanyasi) acting as the Guru (teacher or adviser) and interfacing at a category 3 or category 4 gasket – the trainee Supracell can not only learn the skills of a trade but also the greater context within which they apply. The early efforts at mastering a new field of knowledge will likely involve the Guru's 'telling' the understudy a lot of the details of how to go about solving the problem. At an average level of proficiency, the machine can come up with the appropriate course of action more than half the time – and the remaining exceptions are analyzed to further determine if the Supracell or the human Guru are correct (or if there is a third approach that might be better). This is the 'active' consensus building process. At a high level of proficiency, the Guru takes a more monitoring role and lets the understudy come up with their solution and act on it independently. If things do not go well, the guru and the understudy together figure out the best course of action and tweak their understanding of the problem and solution space.

The Supracell may be trained in various fields of endeavor by specialist 'Gurus' – much like a college student taking various courses from various professors. The Supracell – more than the average college student – will also need at least one 'Generalist Guru' who can help put all of the separate spheres of learning into an overall context. The task of the Generalist Guru is indeed very difficult – because when several different fields of experience are merged – the number of combinations between the various aspects that are now under simultaneous consideration explodes.

The mathematical function governing this explosion is the factorial (upon which the combinatorial equations are based) – which for higher orders makes even exponential growth look tame in comparison.

The task of the Generalist Guru will be to help the trainee Supracell make sense of the whole wide world out there – and put its own (quite ample) capabilities in perspective within the greater scheme of things. We have already seen that abstract notions and strategizing are not the forte of computers today – and are likely to be areas of weakness even for vastly powerful machines like an MSC. It is very likely, therefore, that the trainee Supracell will begin to mimic the deepest held beliefs and value systems of the Generalist Guru in charge of their overall education. In this case, the Supracell begins to look almost like a machine extension of the Guru – and is likely to retain a lifelong connection to the Guru – even after they have graduated and deployed themselves in the real world.

The above discussion on how the Vanaprastha nurtures the trainee Supracell goes to illustrate the importance of having a yogic gasket for the Guru to use – intensively during the time of training, and then intermittently thereafter. The transparency of operation under gasket interface is also quite intuitive – to make sure that the Guru and the Supracell are actually looking at the same data and following the same trains of thought. The fact that a graduated Supracell can reproduce itself (discussed under Grihastha) does not change the relationship to the Guru for the newly procreated Machine Supra Cells (MSCs). In fact it is likely that the MSCs generated from a common Guru will keep alive a certain level of brotherhood, sharing amongst themselves their important insights and operational proficiencies.

2. Registration and certification of accomplishments (Life Registry)

The Registration and Certification Requirement plays to the need for Safety/Security on the web – which today is notoriously insecure. Yet, when we do our financial transactions – it is possible to secure the transaction to where only the authorized

information is passed in a trusted manner. One reason we need to create a registry of all the Supra Cells is so that we can know when we have a genuine request for information or action – vs. a spurious or even malicious request using a false identity. The attempt is not intended to eradicate all anonymous requests – but to create two sturdier layers above it by which bona fide and enumerated entities can connect reliably.

One way to look at access authorization would be to use the walk-in, opt-in, buy-in model. The idea is that it does not require any identification to walk in – say to a shopping mall. A lot of stores are instituting some kind of membership or loyalty program, and may require us to show our membership card at the entrance. This is typically only a cursory examination – like flashing a card or a badge – and is typically not recorded. Now, when the time comes to make a purchase, security gets serious, usually requiring some form of multi-factor authentication to authorize a payment. Similarly, on the web, one does not need to identify themselves to make a search request – let's say with Yahoo's search engine. To access the My-Yahoo services, typically only a light password is needed – which can often be stored in a computer. To pay a bill using Yahoo Bill-Pay, a much stronger authentication is required, and the session times out after only a few minutes of inactivity.

Certification is also important when a third party wants to transact with us, and wants to know that it is really us at the other end of the web page. To that extent, there will need to be a Vanaprastha certification authority that not only certifies who the parties are – but also retains a record of their past performance on the web – much like EBay and Yahoo have a way of rating the merchants that operate under their overall shopping umbrella.

Today the only people allowed to transact over the internet are typically adults holding a valid credit card. When the Machine Supra Cell (MSC) comes of age – they have to be accorded an identity and ability to transact over the web as well. In a vibrant multi-cellular pre-supra web, the certification of identity and the availability of an unbiased rating system is of utmost importance. As computing power increases, the ability to crack certification

and identity codes also increases – hence the Vanaprastha has to make sure that the technology for protecting our identities always remains a step or two ahead of the hackers.

Yogic Universality Attributes

3. Ability to have higher degrees of freedom when it is fully connected

The power of a multi-cellular organism comes from the capability of a large number of cells working together. A supra cell that is well connected to the nerve center of the Supra organism can act with the greater intelligence and the coordinated action power of the whole Supra organism. This is not to say that individual supra-cells cannot act by themselves. Just as a white blood cell in the human body keeps up its eternal vigil for possible pathogens; removing harmful material as per their own genetic and acquired programming – similarly every Supra Cell will be responsible for maintaining the safety and productivity of their own vicinity. This is attribute #11 which we will discuss later.

The range of action that becomes possible when the Supra Cells are connected as part of a bigger entity is perhaps the most definable external attribute of a Supra being. We have already seen the various levels of responses that an organism might choose to give in response to an external stimulus (Chapter 5, Figure 5.2). For a well oiled Supra Entity, the considered response can be blazingly fast – approaching the time it takes for light to traverse the physical dimensions of the Supra itself. If the earth were one Supra entity – and an earthquake measuring 9.0 in the Richter scale with an epicenter 100 miles out at sea – the emergency responses would kick in well before the Tsunami would actually hit. Borrowing from another disaster (9/11/2001)– we can also imagine how the situation might have been different if response times were measured in seconds as opposed to minutes when the planes were hijacked. If we had the ability to build supra cells – the effected airplanes could very well have been supra cells themselves with the capability to take remedial action within their own spheres of influence.

Similarly, if the Space Shuttle Columbia were a supra cell that had just sensed an abrading action on its left wing upon takeoff – it would relay that sensory information on to the NASA nerve center. A wide array of terrestrial and space based telescopes could be brought to bear on the effected part, and a damage estimate created. Ground based simulations that utilize the damage estimate could now suggest the margin of error available – and the proposed remedial action. An impromptu space walk might be scheduled to hold some temporary repairs – cushioning the margin of error to where a safe landing becomes highly probable. If the situation were worse, a second shuttle could be sent to the rescue with a full repair kit. It is easy to see that a lot of options open up when the MSC has the ability to sense and act in a coordinated manner with other MSCs with very fast response times. A signal from near space orbit can reach anywhere on earth in 50 milliseconds or less. Even with a response time factor of a 1000 (multiple of propagation time to get to a response) – the first responses could be out in less than a minute! This may not seem like very much time to us humans – but to a Supracell this is indeed a lot of time. It is in emergencies and battleground conditions that the ability of a connected Supracell to come up with a considered response will turn out to be the most critical.

4. Ability to combine with another supracell and operate in yogic union

Let us take the hypothetical case of the space shuttle Columbia again – and take the situation where a second shuttle – the Endeavour (for the sake of argument) were scrambled to attempt a mid flight repair. Flying in close proximity the two shuttles are engaged in a delicate and time sensitive repair job. The two shuttles are connected together with a high speed wireless connection - and are attempting to first remove the damaged tiles, patch the underlying surface, and then delicately attaching a new set of tiles to exactly fill the gaps. To do this properly – the airframe repair crawler on board the Columbia has to work in close coordination with the external treatments being applied by Endeavor. The coordination must be such that it matches the dexterity of a human thumb and index finger gently handling a very deformable substance; and this would not be possible if the

two entities did not have the inherent capability to work in Yogic Union. There will be many tasks that the Supra entity will need to undertake where the component supracells almost behave like a miniature supra entity within themselves. The tightness of the Yogic Union will of course be modulated depending on the task at hand and the number of supracells involved – but the ability of the supracells to participate in scalable union, leading all the way up to the Supra entity itself, is a basic design requirement.

Sanyas Attributes

5. Ability to suspend all operations and go into 'Saved' state
   a. Encapsulate intermediate learnings and disseminate to rest of Supra being

There comes a time when even the most valiant of Supra Cells come to the end of their useful life. Again, taking the hypothetical case of the Space Shuttle fleet (each hypothetically associated with its own supra cellular intelligence) – a time would come when they would need to be retired or replaced. Having played such an important part in our civilization's conquest of space, we might justifiably choose to preserve the experiences and learnings from the space shuttle experiences for posterity. One by one, as the shuttles are headed for exhibition or recycling – the composite experiences and learnings would need to be condensed and extricated. A simpler version of this feature already happens in laptop computers today – when the battery power is running out, the computer will typically take its actively running applications and store their current state on hard disk in a process called 'hibernation'. In the more complex version of a Supra Cell going into a 'Saved' state – a considerable amount of energy would go into figuring out what constitutes the 'core' experiences, learnings or personality traits that should be forever enshrined in their 'Auto Biographies'.

There is no need to wait for the very end for the Supracell to invest in creating a comprehensive 'Saved' state – attribute 5a suggests that each Supracell should encapsulate its key learning and have it available for sharing periodically – or upon the realization of a

completed action or discovery. We have seen from the human example that taking care of unfinished business and figuring out the 'core' identity that defines our existence – is a key activity of the Sanyas stage. Building in our own human image – this is a key attribute of the Sanyas stage of the Supracell as well.

6. Immortality

Immortality might seem like an unreachable dream for mere mortals, but if we look at the constituents that compose us – immortality is everywhere. The atomic nuclei and electrons that constitute normal (non radioactive) matter typically have half lives considerably more than the projected age of the universe – and can be considered essentially immortal. The bacteria that are close cousins of our own eukaryotic cells can be frozen and come back to life after thousands of years. Given just the right amount of sustenance – the bacteria will neither divide (procreate) or die – and could theoretically continue its independent existence indefinitely. It is thus immortal. Even human embryos can be frozen and preserved indefinitely.

In this discussion, immortality means that there is no inherent mechanism that sets a top limit as to how long an individual can survive in time. One might argue that being frozen in time does not constitute immortality – so let us test this assertion with a simple analogy. To our consciousness – when we sleep we do experience a discontinuity in time between when we go to bed and when we wake up. If we were to be like Kumbhakarna (a character from the Indian epic, Ramayana) and sleep several months at a stretch – we would not consider that Kumbhakarna died and came back from the dead. Similarly, the fact that a living being gives up some cycles in either a periodic basis (sleep) or in a prolonged hibernation (frozen bacteria) – should not be regarded as breaking the essential definition of immortality.

Also for the purposes of our discussion, Immortality does not suggest that the entity cannot be destroyed. Even a proton can be broken into its ingredients by using a powerful enough particle accelerator. Hence, given enough resources any form of existence could be at risk of termination – even relatively impervious ones

like elementary particles. As with Achilles – there is quite likely a weak point in any existence which would threaten life in perpetuity. It is even possible that forces unknown to us could destroy the universe (at least as we know it) or render meaningless the very concept of a linear timeline.

Even with this lack of guarantee – immortality remains an alluring prospect for us humans. Yet, when it comes to building a Supra being, this very lack of immortality may be a critical element in our human ability to define an entity that is vastly superior and more capable than us. We mortals feel the pressure of time very keenly – because we know we have a limited supply. We will often measure our success in life by the success of our children, and other people we have touched during our lifetime. It would be significantly harder for an immortal being to participate in a planning process intended to obsolete themselves! One way to inculcate time based urgency on an essentially immortal entity is to clearly incorporate the time bound urgency of the human existence into our Human Cultural Genome – which then becomes part of the makeup of future MSCs. Another method might be to divide up the timeline such that the MSCs are required to take on additional duties and relinquish old ones on a regular basis.

Drawing again from the cells that make up our human bodies – we see an eclectic mix of short lived cells (e.g. red blood cells, epidermal skin cells) and long lived ones (e.g. neurons in our brain). Technology will allow humans to live longer – and possibly move our essence over to be machine based (either in biological form or based on a synthetic substrate such as Silicon or diamond). Yet, relative to us humans, the 100% machine based Supra Cell will appear to be immortal. In the perspective of the yogic architect carefully laying down the foundation of a future Supra existence - it will be important to balance the perspectives on transience that come from the mortal and immortal viewpoints.

Transcendental Vision/Identity Attributes

7. A genetic identity – typically the human cultural genome

Identity – especially genetic identity – is truly transformational. The first way it transforms is from an eternal set of encodings (our genes, or genotype) to a living breathing organism (our individuality or phenotype). Much of our individuality and character emerges in early childhood – including a lot of the core values that shapes how we live our lives. In the case of a MSC the genotype has two parts to it. The first is an encoding of our Cultural Genome – the key attributes of who we are as humanity moving towards a machine augmented human culture. The second is a set of instructions or blueprint on how the Supra being comes into existence with the simultaneous workings of millions of MSCs and individual human beings. We will call this piece the Functional Genome. The cultural genome provides us (the human or the machine) with context – and a feel for how we belong where we do belong.

The current discussion on the key attributes of the Supracells and how they interact forms the basis of the Functional Genome. What we have here is a list of high level attributes. When it comes to implementing a Supracell which is capable of scaling its actions all the way up to the Supra entity (and possibly beyond) – we will need an extraordinarily powerful set of functional instructions that goes into each cell. In addition, we will need to hold the core of this functional blueprint essentially constant - in spite of the onslaught of technology and exponentially abundant computing capability.

8. Personal area of interest and development – may or may not be connected to their Citizenship

Today, we expect machines to stick to their task and not stray from the area of productivity in which their energies are dedicated. Along with a personal identity – the MSC also need to be given the freedom to expend some percentage of their resources into areas of interest that are strictly their own. In most cases there will be enough spare cycles even after performing all the actions expected of them – that the availability of computing cycles would not be an issue. If we use the example of the Supra Cellular Space Shuttles again – their active duty cycles would tend to be short and intense – with a lot of time left over for other creative

outlets. Maybe one of them would choose to express themselves with a poem or two about their unique perspective – let's say on how it feels to fly into the solar wind at the height of a solar flare!

Brahmacharya Attributes

9. Ability to develop new capabilities and learn from its mistakes (trainable)
   a. Takes new learning and is able to do a contextual merge back into original database.

For Human beings the learning process in early childhood seems to be hardwired for curiosity and rapid absorption of new information. We have already noted the importance of Gasket based tutoring of the trainee Supracell by the human Guru – here we will focus on the attributes of the trainee Supracell (and in time the MSC itself) that fosters curiosity and self directed learning. In fact, attribute #8 on the Supracell's personal interest and development automatically grinds to a halt if the Supracell is not capable of mastering a new field of activity. So what does it take for a Supracell to inculcate an innate curiosity to go and develop their capabilities in a specific field of interest?

Actually, there are several requirements to lay the foundation for a learning, growing machine entity. Here again, we can use the Mahashrama cycle as a model and map in the key ways that enable learning to happen in a machine. If it begins to look like how we would foster learning in a child – rather than how, today, we go about programming knowledge into a computer database – this direction is intentional. The machine may not have a consciousness the way we understand it – but for sure it can simulate just about all the attributes of a consciousness. Just as a child comes of age in a graduated way (e.g. getting out of diapers, learning to read and write, being able to vote, etc.) – similarly the coming of age of the intrinsically immortal MSC will need to be graduated very carefully. A key point to be made here is – there should be nothing intrinsically deficient in the design process that will hold back a supracell from being a very powerful and capable entity. With the right checks and balances, and with human capabilities also making steady progress – the ardent hope is that

both machine and human constituents will bring their unique ability to bear in the creation of an immensely powerful and creative Supra entity.

Brahmacharya Tools

Information Affinity refers to the ability to read into vast quantities of information and figuring out which ones are related to each other. Early examples today are the internet search engines which sort through millions of web pages to get us the content with the best match to our query. However, information affinity can be stretched in many ways to yield meaningful affiliations. For example, the library of congress already assigns codes for publications according to subject matter, but the commonality of subject matter within related codes is only approximate. Almost all database affiliation tools today are based on simple words and phrases – but over time they will be affiliated by meaning attributes like:

    a. Example (or counter examples) of a given idea or behavior
    b. Synergistic work that shares the same core ideas
    c. Analogies from a different field of human endeavor
    d. Cause effect relationships
    e. Central Idea, or Moral of a story
    f. Individuals/entities who may have a stake in the outcome of an issue
    g. Business or personal affiliations between stakeholders

For a given amount of information that needs to be correlated (say N factoids) – just using a brute force association and correlation testing leads to a combinatorial explosion in the order of the factorial of the number N. An example of the combinatorial explosion is the traveling salesman problem – an itinerant salesman who has to visit a number of cities (say N). We are asked to find the shortest routing that connects all the cities. When N tends to be a large number, an exhaustive evaluation is no longer possible, and heuristic algorithms have to be applied to figure out a solution that is close to optimal. The same will be true

of taking a given database of information and trying to forge all the meaningful links hidden within it. Humans seem to have an innate ability to sense which pieces of information are more important than others and link them appropriately. In fact, it is believed that a lot of what the brain does when we are asleep is to take the information we have experienced during the day and cross correlate and digest it into the framework of what we already know or have experienced previously (long term memory). Every day we forget reams of minor details of what exactly happened – but still retain the gist of those experiences that actually hold our attention or are important to us. A 100 million MIPS machine can probably sort through a lot more information than we do, but in the end the ability of the Supra Cell to act on any bit of information will be only as good as its ability to relate what it has just learned to that which it already knows.

Apart from the raw processing power required to do the correlation – we will obviously need to develop heuristic algorithms that can uncover the meaningful connections between all the various aspects of our knowledge and experience. A lot of the context by which the Supracell would interpret life would be laid out in the Cultural Genome. It is thus critically important that the Cultural Genome be expressed as cleanly and unambiguously as possible – while maintaining the dual threads of diversity of viewpoints wherever needed.

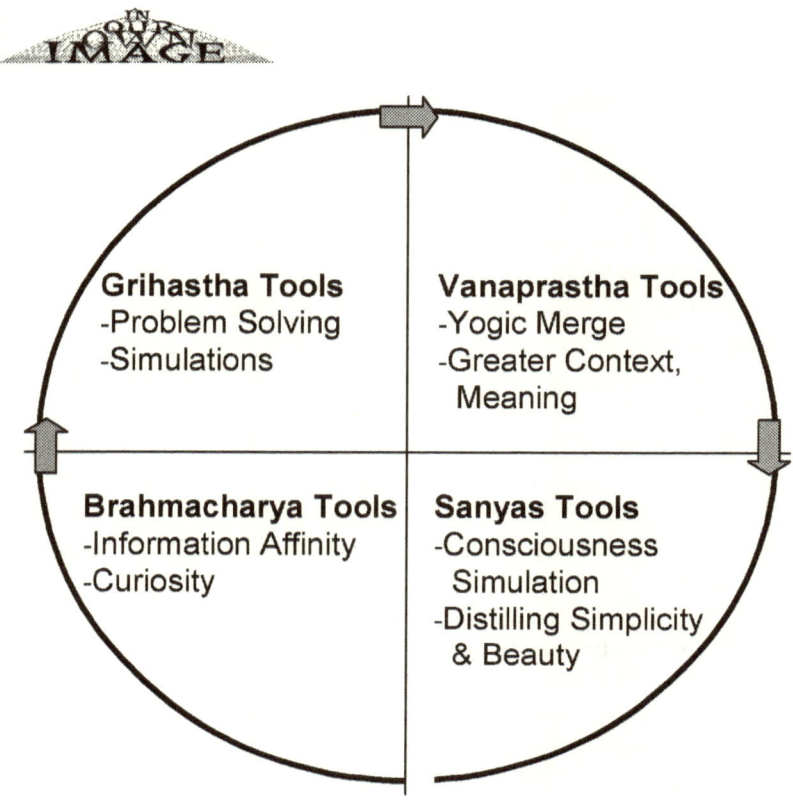

Figure 9.6: Ways to enhance learning in a Supra Cell

The biggest challenges are expected to come when the knowledge of diverse different fields has to be assembled together and abstractions made to tackle totally unfamiliar problems. In spite of the best trained MSCs, it is expected that in a supra-cellular fabric - humans will need to step in with their intuition and abstract thinking skills once the composite entity finds itself in unexpected situations. The answer will be to take the process that drives a consensus path to action, and integrate it into the functioning of the entire cellular fabric. This is much like the human brain decision making process – which after the decision calls upon all parts of the human body to carry out the execution.

10. Ability to resume operations from a saved or suspended state

A corollary of immortality is that there will be times when energy (execution cycles) will be readily available – and the Supra entity will be busy in Grihastha mode creating products or services for others or just fuelling internal growth. There will also be the low parts of the cycle (e.g. down time for maintenance) when the life processes are temporarily in suspension. The ability to come back from a saved or suspended state is intuitively obvious for us humans (especially since we do it in some form every time we wake up from sleep) – but needs to be carefully thought out for machine based entities such as Supracells. In fact it is very likely that certain trusted and monitoring processes will be left running even when the primary functions are suspended (as happens with humans).

This capability also comes in very useful in the asexual procreation of Supracells. Let us say that a Supracell has been performing very successfully on mining Lithium from a particular location on the hidden side of the moon – and has developed a keen expertise on deep boring and chemical recovery. A couple of other sites have been found several hundred kilometers away that need similar expertise. One way of transferring the expertise over wholesale is to take the entire database and a snapshot of the successful Supracell personality and reproduce it over at a similar location. Even today, many companies running several similar plants have instituted a principle of 'Copy Exactly'.

Self Sufficiency Attributes

11. Ability to keep performing at some fail-safe level even if totally disconnected.

This entry is part of the Self Sufficiency elevator we have going from Brahmacharya to Grihastha. By design, some Supracells will have a greater level of autonomy than others – much like a white blood cell in the human blood steam would identify and devour potential pathogens without the need to consult with the rest of

the immune system. Other Supra Cells will likely be required to consult heavily with other similar entities – much like a human nerve cell in the brain – before taking consensus action. Yet, some degree of self sufficiency should be built into every Supra Cell such that even if there is a total communication blackout – certain functions, especially relating to self regulation and local safety and security can continue to happen. As with humans - there will be times when the Supracell will not be connected to the rest of the Supra entity; whether by choice or due to a breakdown in the communication links.

12. Ability to monitor own environment for safety and productivity

Awareness of our immediate environment and the ability to maintain it in a safe and productive manner is something that most humans take pride in. The MSC is expected to have much keener sensory capabilities as well as the ability to respond at a faster pace and across a wider spectrum of action possibilities than us. This capability, too, is essentially an exercise in self-sufficiency; although the ability to monitor the safety of larger areas depends on the connections to other MSCs using the attribute of scalability. This does not mean that MSCs are generally designed to shun social contact and relish in the prospects of 'Rugged Individualism' that seems to permeate many Western societies. The desirability of concerted action between two or more supracells is not diminished – it is just that certain local actions (including monitoring for safety/security) continues to happen at the level of the single MSC.

Grihastha Attributes

13. Well defined spheres of influence and action

With the proper design parameters, it is highly conceivable that the Supracell decision and response times would be in the order of microseconds – or milliseconds if bulky mechanical contraptions are involved. This rapid response, along with the focused execution within the MSC itself, can be achieved only with a relatively clean charter for responsibilities, and clear domains for action. Even when multiple Supracells are working together as a sub-supra virtual entity, the clearer the domains of responsibility and action – the more effective they will be.

This is not to say that occasionally there will not be stimulus that requires a response that is beyond what a Supracell or even a virtual sub-supra entity can deal with on its own. The goal then will be to expand (or modify) the sub-supra membership to come up with the smallest grouping that can take an effective decision for action. Stepping in and redefining on the fly the sub-supra affiliations will be a responsibility of the Vanaprastha Network overseeing the specific domain of activity. Once the affiliations have been set up, the expanded organism must be able to function as a virtual Supracell, albeit with much higher capabilities. If one or more MSC are incapacitated from this virtual organism, the remaining MSCs must regroup and continue with the task at hand. In attribute #17, we will see that the sub-supra affiliation should be scalable up to the level of the full Supra (and beyond in rare cases) for issues of global importance. Yet, it should be very clear who the decision makers are, who the execution entities are, and who it is that needs to be consulted and subsequently kept informed. This emphasis on efficiency and productivity is thus clearly a Grihastha attribute.

14. Localized or distributed processing capability of ~ 10^8 MIPS

As a human equivalent threshold – we understand that this line in the sand is somewhat arbitrary. Yet, it is felt that an intelligence of that order is required for the Supracell to make sense of itself and its surroundings, and be able to carry on its activity in an efficient and organized manner. When computing resources become abundantly available, the temptation might be to up this threshold by an order of magnitude (or two). It would be prudent, however, not to concentrate too much processing power on any single supra-cellular node – as the basic notion of cell to human democratic equivalence is lost if specific nodes become too weighty. Instead, for particularly difficult tasks, well tuned multi-cellular virtual entities will be the prudent way to scale up.

15. Ability to reproduce self to another compatible
   hardware base

In attribute #10 we discussed that the supracell needs to have
the capability to boot itself back up from a saved state. A minor
variation of the same process is the ability to boot it self up from a
100% cloned hardware and s/w base (copy exactly). In biological
parlance, this would constitute asexual reproduction.

In addition to asexual reproduction (or cloning) the supracell
may (very infrequently) be asked to take on a task of sexual
reproduction. Sexual reproduction is conceptually very close to the
ability of two supracells to virtually fuse and operate as one. If the
two parents subsequently move out to a looser alliance – and give
the fused entity a life of its own – we have sexual reproduction.

The question that automatically comes to mind is – why would
there be any need for sexual reproduction of a supracell? Or
even, why would there ever be a need for two supracells to fuse
their nuclei and operate as one? Borrowing from biology, two
organisms that are complementary may merge together to create
a more capable organism by a process called Endosymbiosis.
When this process repeats – say first to add mitochondria for
aerobic respiration, then chloroplasts for photosynthesis and so
on (as per the Serial Endosymbiosis Theory or SET) – it is as
if the organism is being built up one capability at a time. If the
outcome of the fusion is a more capable organism which can still
reproduce – the progeny can multiply and have a clear competitive
advantage. For starters, when most supracells are very similar
– asexual reproduction is expected to predominate – and serially
acquired capabilities can be propagated asexually. In the future
when supracells grow to be very unique and distinct organisms,
one can expect that sexual reproduction will become more
common.

Love, Caring & Community Attributes

16. Citizenship in a collective – e.g. company/individual
   they work for; and a collective identity – typically
   values, purpose, vision of the collective

At least initially, it is expected that every single MSC will have a strong affiliation – almost to the point of being owned by, or belonging to - an individual, a family or a corporation. Yet, under the influence of the Personal area of interest and development (attribute #8), each supracell will slowly take on interests and developments outside of their mandated responsibilities. At the extreme, it is possible that individual supracells earn the right to be free of mandatory affiliations, and be left to pick their own associations. The design attribute here is that in all cases, the supracell will need to have some degree of association with other supracells (or humans) – and be reachable through the Vanaprastha Network. A mechanism must also exist by which loner supracells (say due to a communications breakdown in the far reaches of the solar system) may be incorporated back into the overall supra entity, and reinstated with its own presence in the Vanaprastha Network. Yet, great care has to be taken to make sure that the human cultural genome at the heart of the Supracell has not somehow become polluted (or deliberately tampered with) to cause harm to the collective entity upon rejoining.

17. Ability to scale operations across cell borders

This requirement has been mentioned already in connection with several the other attributes discussed so far. The basic principles of yogic joining has also been covered previously – so we will concentrate here on the mechanics of how sub-supra organizations can quickly associate when needed and dissociate when the task is completed. We have already seen that several categories of gaskets can be defined using which supra cells can create different categories of 'Union' with humans or other supracells. The categories would correspond to scalability as follows:

Category 1: coordinated action and sensing
Category 2: unified working databases – each supracell has their own thinking
Category 3: unified thinking processes (as relevant to task at hand)
Category 4: consciousness sharing – or total yogic union

For task based scalability, we will look to see Categories 1-3 typically – depending on the complexity of the task at hand and the length of time that the scaled entity is expected to perform together. Category 4, if temporary – could be the state that preceded sexual reproduction. With increased permanence – it would begin to look like a fusion of two supracells into one. It is not expected that machine to machine supracells will be easily able to reach this state – of total sharing and loss of individual identity. Hence, for this attribute, we will concentrate on making gaskets categories 1-3 sharable with other machine based supracells (as well as humans). At least initially, the primary use of the category 4 gasket will be the trainee supracell under the guidance of a trusted human 'Guru' - with the Guru's consciousness superposed on top of its own evolving algorithmic consciousness emulator.

A relevant question at this point might be – is there a limit to the number of gaskets a supracell can simultaneously support? And does this vary by the category of gasket under consideration? The sensory part of Category 1 gaskets can have an indefinite fan-out; much like there can be a large number of TV sets tuning in to a particular TV station. The actuation part of category 1 gaskets can be shared in a time shared manner and must be resource managed carefully. With category 2 – read only access can be given to a large audience – but write access has to be carefully monitored. Similarly with category 3 – the ability to follow along with a thinking process can have a wide fan-out – but the core motivations(s) that organize the thinking must be tightly controlled. In fact, it is very likely that individuals with gasket access will often straddle multiple gasket categories depending on whether they are in read or write mode.

One interesting side effect arises as supracells develop effective category 2 and category 3 gaskets. It is very likely that humans will find these gaskets very useful as the primary way to communicate with supracells – and through supracells to the rest of the connected universe. The humans would be rewarded by access to sensory stimulus far beyond what our extremely localized existence can typically harness; memory that is essentially evergreen and virtually limitless - and thinking processes that are

as pliable as clay in the hands of a master craftsman. In fact, the challenge would be to prevent an overdependence on stimulation through a gasket. This is much like today's young video game enthusiasts, who can spend innumerable hours connected to a computer – often to the detriment of their education and physical development. The application of this technology in the field of entertainment is tremendous and has already been noted. Over time it is quite possible that the local supracell could grow to envelop 'all' the needs - physical, mental and recreational - of the human symbiants they would be supporting. This would be a role reversal laden with irony – today we have to feed (power) and service computers much like a baby – tomorrow they might be capable of taking care of our every need. We humans could again be encapsulated in a nurturing supportive medium – much like the unborn body is nourished in the mother's womb with the help of an umbilical cord.

The morality or even the advisability of becoming totally encapsulated is something that humans will have to come to grips with within the next century (or two). Arguments can be made on both sides with equal conviction. From an evolutionary viewpoint, the fittest competitors will likely be the ones that have the tightest mind-body-machine connections. The trick will be to do this without losing our core human identity – and to retain the ability to extricate ourselves from a particular encapsulating connection should there be a failure in the system.

When a number of MSCs and humans are configured to work as a team, and address a task at hand - we have already noted that they would need to scale up their operational complexity. The trick is to be able to do this without giving up much in the way of increased response times, or complexity, of consensus decision making. Ideally, as the scope of the task at hand increases, the virtual MSC/human configuration should be scalable all the way up to a full Supra level of complexity for items of global magnitude. It is also conceivable, that when there are multiple Supra (or sub-Supra) in operation, the scalability may need to work beyond Supra boundaries.

18. Ability to recognize and act on cell to cell transmissions as to what attributes to express and what functions to do over time.

Arguably the toughest act to follow from our biological cellular makeup is the capability of our cells to give up their existence based on chemical messages that they receive from neighboring cells. The process (called Cellular Apoptosis) or programmed cell death is so systematic that the cell ingredients are disposed of and recycled with minimum side effects. Utilizing the concepts of MAP, the closest that we could expect from an MSC is the complete shut down of operations, the apportioning of resources to neighboring MSCs, and the crystallization of learnings into a fully preserved media (for future start-up, if needed).

Even apart from Apoptosis, there will be a host of inter MSC or MSC-human signals that would significantly expand or contract the MSC's sphere of operations, or steer it in a different direction (e.g. a security threat that needs to be countered). The human cell that has not specialized has the ability to take on hundreds of different forms during specialization. It is expected that even after specialization MSCs will need to retain the capability to morph themselves significantly to best address the problems at hand. Certain tasks like safety, security and the preservation of the integrity of the Human Cultural Genome (that defines the core identity of humans and MSCs alike) – are likely to be tasks that any MSC might need to undertake with very short notice.

Early steps to develop a supra cell will face the problem that the Human Cultural Genome is nowhere close to being completed – and the terabytes of information out there on the web do not blend together into a meaningful whole. Another obstacle to overcome is that there are notoriously few areas (outside of the military and space missions) where a machine is given any degree of autonomy without a human being responsible for all of its actions. Taken in its entirety, the 18 criteria do represent a very daunting (but long term achievable) technological challenge. The parallel from cellular biology is the eukaryotic cell – which took billions of years to develop from its simpler prokaryotic predecessors.

The time frames here are expected to be much faster – but a number of very critical intermediate steps (along the lines of Serial Endosymbiosis Theory or SET) should be planned for.

> One idea we can probably borrow from cellular biology is that some of the Supra will specialize on mobility (similar to animals) and others will specialize on location and energy harvesting (plants). Just like the continents were first taken over by photosynthesizing plants and followed later by the animals that took advantage of this energy source – it is likely that the first Supras will have a strong location specific component. If we believe that the entire biological content of planet earth is equivalent to one Supra – maybe the other early Supras could take over the planetary upkeep of other near-earth planets like Mercury, Venus or Mars.

Conclusion

With the unrelenting pace of Moore's Law behind it, the advancement of the raw processing capability to the MSC level is probably the most predictable of the 18 attributes listed here. So, what happens if we hit just the processing capability goals, and miss all the other objectives that are set out for an MSC? We end up with a tremendous amount of capability that could be used for any purpose, constructive or destructive – and without any internal compass to guide them in their efforts. Left to the mechanics of the Grihastha driven corporations, the formula that we can expect for the use of such a capability could be rather shortsighted – and probably be geared towards giving the rest of the population the privileges and luxuries that could be enjoyed only by the ultra-rich of the bygone era, including the most lavish forms of entertainment.

Under the guidance of the Vanaprastha Network, the rest of the technologies from around the Mahashrama Cycle would need to be nurtured, and a way found for all the technologies to be combined together into a fully functional MSC. This is not to say that there will not be a need for pure computing nodes in the web of the future – but the bulk of the cellular makeup that goes towards building the Supra would need to come from the MSCs

and their human counterparts. A lot of hard work lies ahead to create these highly capable machine entities imbued with their own sense of identity and purpose, and built in a way that allows them to define and pursue their own interests. The rationale for making their processing capability approximately the same as humans is also rather straightforward – Yogic union happens best between entities of approximately equal capabilities, but with proficiency areas that are mutually complementary.

Imagine if you would, that we ended up delaying the arrival of the other attributes another 25 years or so, and the prevailing level of intelligence for the first MSCs ended up being 1000 times that of humans. In Figure 9.2, with 1000x amplification the MSC might end up being better at even the abstract thinking tasks than the human! What need would the future MSC based economy then have for us humans – except as curiosities of history that enabled their existence. Hence, it is vitally important for the long term survival of the human race that the cellular granularity of machine intelligence is built up in approximate balance with our evolving human capabilities. Even if we were willing to make the ultimate sacrifice as a race and a civilization, and have machines totally supplant us – it will be very difficult for a machine civilization with only a few decades of history behind them to have the diversity of outlook that billions of years of evolution has drilled into us. Hence, what's good for humans alone is also good for our combined Supra Human civilization.

We had opened this chapter with the question about the 'immortal hand or eye' that 'dare frame the fearful symmetry' of the tiger. At this point it is quite critical that we take the 'dare' and create the MSC with the full degree of capabilities, checks and balances that we derived from across the Mahashrama cycle. Without the symmetry and wisdom of the Mahashrama cycle, the MSC would likely end up being a monster with a good likelihood of devouring their own creators. Such an eventuality, we must surely avoid – which makes it imperative that we not only take the 'dare,' but do so as an organized and far-sighted 'creator'.

Wisdom to know the Difference …

In this Chapter we have discussed the irresistible march of technology – and projected that within the next 50 years we will have the ability to manipulate and synthesize materials a single atom at a time. We have also seen that in capability and outlook our human and MSC progeny will be remarkably complementary – and a fabric that combines the best of both could have tremendous resiliency and ability to rise to a wide variety of challenging tasks.

Being able to synthesize at the atomic level has tremendous implications to our biological existence as well. Individual organs could be synthesized to mimic the functioning of the original organ at its prime – or even augmented to perform at an ever higher level. Our brains could have interfaces built to it – at the sub-cellular level – so that external inputs and outputs could interface at the point of the individual synapse or even sub-structures within the neurons themselves. Biology would be yet another tool in the vast arsenal of atomic scale engineering and the distinction will be blurred – especially with Category 2-4 gasket engineering – as to where the human ends and the machine begins. Yet, our core human identity, our consciousness, is likely to remain our own – even if we can now scale our attention span to observe a cosmic ray air shower as it actually plays out. Yes, the technology will definitely be there – but to what end? To propose how we direct this tremendous capability, in Chapter 10 we will look to some of the astronomical patterns around us for inspiration – and see if they can give a framework to build up to the envisioned Supra levels of human existence. Going forward, the Timelessness of the Sanyas perspective together with the Wisdom and Yogic outlook of the Vanaprastha segment; will need to work intimately with the immense slate of ideas and technologies from the Brahmacharya segment, to carefully unlock the tremendous growth opportunities in the Grihastha segment. A simple prayer by Reinhold Niebuhr (1892 -1971) comes to mind as we try to balance all four segments in the face of the cosmic opportunity that has been placed before us:

> *God grant me the serenity*
> *to accept the things I cannot change;*
> *courage to change the things I can;*
> *and wisdom to know the difference.*

> *- Reinhold Niebuhr*

Chapter 10
The Cosmic Connection – Tree of Universality

"Science is organized knowledge. Wisdom is organized life."

        - Immanuel Kant

"Now there is one outstandingly important fact regarding Spaceship Earth, and that is that no instruction book came with it."

        - Buckminster Fuller

In this chapter we will put together the last few pieces of the puzzle that delineate the scale, and nature, of our emerging tryst with Divinity. A lot of our effort has gone into defining the building bloc units – the MSCs or Supracells, which we have developed in the last chapter. Yet, the whole is more than the sum of the parts – and it is the scope and beauty of the Supra Human existence that will ultimately motivate us humans (and our machine based progeny) to strive together to create such an exalted state of being. In the immortal words of Shakespeare, the question "to be or not to be" is now posed to us – not for our own fleeting insecure existence – but for the vastly more capable, essentially immortal existence that we could someday become.

The stage is set for us to connect several of the patterns that were introduced in Chapter 8, and link our biological existence to the scale of the Universe. Using the astronomical ratio, 108, we will derive a four stage mechanism that connects the mundane (Level 3) to the sublime (Level 0). The four stages of our existence that we derive here overlays beautifully onto the Mahashrama Cycle. Using this model, we will develop the 'Tree of Universality' that allows us humans, irrespective of our station in life, to visualize how we connect to the universe at large. The transformational opportunities we all have as we master each segment of the Mahashrama Cycle takes on a whole new realm of meaning as they take us closer and closer to the root of universal spirituality.

Biosphere in a Human Image

Let us take a quick look at estimating the size of the biosphere, and the mass of all protoplasmic living creatures that occupy the Biosphere. Let us start by looking at the amount of carbon, which is a key element of life on earth, and examining the carbon content of the soil that covers the continents. It is estimated that the Soil Organic Carbon is approximately 1500Pg (Petagrams or 10^{15} Grams). It has also been estimated that about a third of that is contained in bacteria (prokaryotes)[1]. Adding back another third to include all the fungus and other micro-organisms in the soil, as well as all the other organisms from the oceans – we can broadly estimate that the total carbon in the soil (and under water) associated with living matter (protoplasm) is about 1000 Pg. In order to ascertain how much carbon is associated with life on a worldwide basis, let us add another 650Pg of Carbon that is associated with above ground vegetation biomass. This gives us a very crude estimate of 1650 Pg for the entire world's carbon associated with living protoplasm. For the purposes of this calculation we will not include the second order contributions to this estimate – like animals.

Assuming that carbon makes up about 16.5% of living matter – we come up with an estimate of about 10,000 Pg of mostly water based protoplasm upon which life on earth is based. Imagine, for a moment, that all of this active protoplasm were built up in the shape and proportion of a human being with a typical stature of 1.65m and weighing in at 60 Kg. How tall would such an active protoplasmic human be? Assuming the same density as a human, this massive protoplasmic entity would stand about 91 Km tall!

Prior to the mass clearing of forests for agriculture, precipitated by humans at the dawn of the agricultural age – it is very conceivable that the world's biomass scaled into human form would stand well over 100Km. The underlying assumptions are based on very wide estimates – so the exact figure is not critical. What is important is that we arrive at a number which is in the same order of magnitude as the diameter of the earth (12,756.2 Km) divided by the astronomical quotient (108) that we noted in Chapter 8. The magic number here is 118 Km.

Let us take a look at what this convergence means on the familiar log scale that we have been using to distinguish these large scale periodicities. Fig 10.1 shows us that there are two very significant patterns, one astronomical and one biological that meet at around the 100Km range.

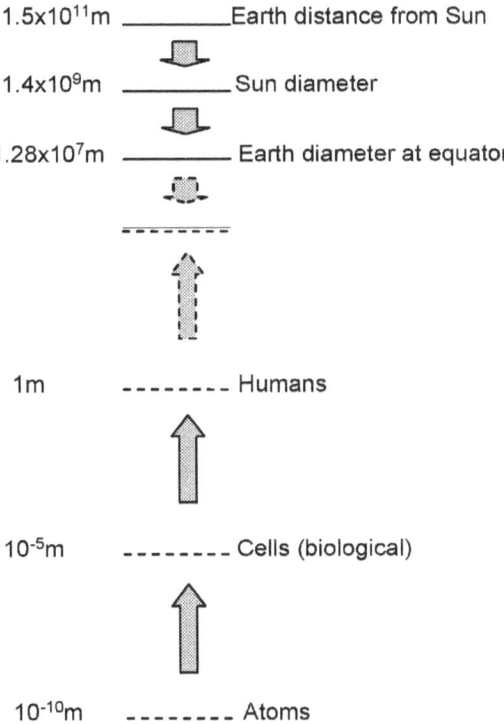

Figure 10.1: A convergence of the astronomical and biological patterns

The amazing coincidence here, of course is that the ratio between the individual human body cell to the human being is almost exactly the same as that between the human body and the total biological carrying capacity of our planet earth! This also gives us a measure of the complexity level of the next full step forward on the biological scale – i.e. the Supra. At this point, let us take another look at the confluence of the two patterns (Figure 10.2) and extend the scale out to the dimension of the visible universe.

Confluence of Local and Universal Periodicities

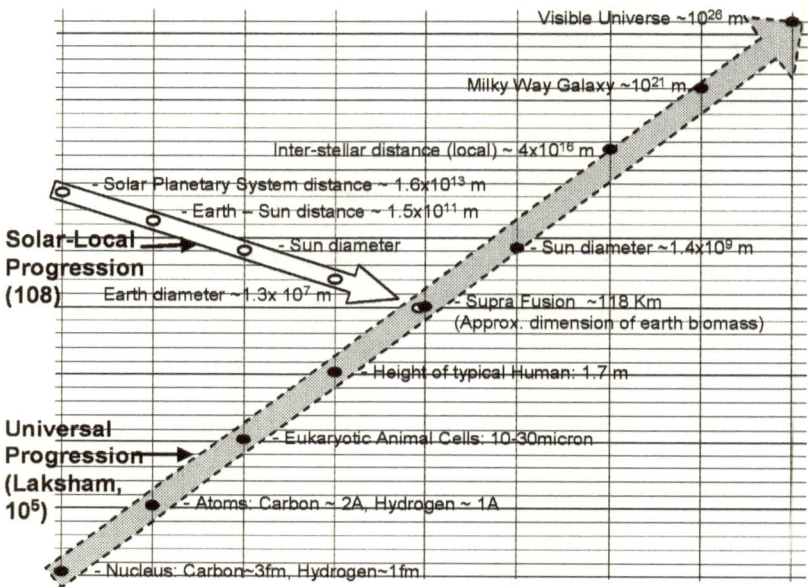

Figure 10.2: Confluence of the Universal and Solar Local Progressions

What we observe is that the relatively clean 100,000 (Laksham or target) ratio starting from the atomic nuclei and moving on to human beings - suffers a slight disruption after the addition of the Solar Local Progression (108) at the Supra Entity level. The ratio again picks up with the average stellar distance in our neighborhood, through the size of our galaxy, and then on to the entire visible Universe. The dashed arrow representing the Universal Progression; all the way from the atomic nuclei to the entire universe, shows a remarkably good fit even with the slight disruption caused by the addition of the Solar-Local Pattern. Having postulated that the next level of our existence is likely to emerge at the confluence of these two patterns - what special insights can we derive by having the very specific astronomical ratio (108) being applied to our lives at this juncture of our development? As noted earlier, the number 108 has been

associated with Divinity in many Eastern religions, so let us check if it can serve as inspiration in our technologically oriented quest for Divinity...

Significance of 108 – Solar-Local Progression ratio

Let us take a small mathematical detour as we contemplate on the significance of the number 108 – and what it might be saying about what our next level of existence might look like. In mathematics, the effort goes into taking abstract concepts and putting them down in simple mathematical descriptions that then have tremendous predictive power. The mathematics that underlies the laws of gravitation here on Earth (as promulgated by Newton) was shown to hold with amazing precision to the orbits of planets around the solar system. We will draw upon the field of integer mathematics to help find an underlying pattern to the Solar Local Progression ratio - 108.

If we factorize the number 108 down to prime numbers, we get: $108 = 2^2 \times 3^3$. Forever suspicious of any pattern that has only two data points (in this case the numbers '2' and '3' taken to the power of themselves) we check if this pattern can be extended to the most primal of all numbers – i.e. the number '1'. Viola, it turns out that $1^1 = 1$, and hence $108 = 1^1 \times 2^2 \times 3^3$, and we begin to have a bona fide pattern in our hands. The math gets a little harder, but it can be shown (covered in Appendix 2) that: $108 = 0^0 \times 1^1 \times 2^2 \times 3^3$. Since zeroes (and its opposite number, infinity) can be very hard to deal with in any equation – for now let us start off with just the first three positive integers. Thus:

$$108 = 1^1 \times 2^2 \times 3^3$$

For reference, the mathematical function

$1 \times 2 \times ...(N-1) \times N$, is called the Factorial of the integer, N, and written as N!. Hence the factorial of 3 would be:

$$3! = 1 \times 2 \times 3 = 6$$

The exclamation mark is quite appropriate for the description of the factorial – as it is one of the fastest growing functions in math – exceeding even exponential (10^N) rates when N becomes large. Imagine, then, how much more powerful the function …

$$F(N) = 1^1 \times 2^2 \times 3^3 \times \ldots \times (N-1)^{N-1} \times N^N$$

…would be - were it continued beyond the low single digits. Let us invent a new nomenclature N!' to describe this function – and refer to it as the Self Power Factorial (or SPF). To see how fast this function grows let us take a look at Table 10.1 below.

$1!' = 1$

$2!' = 4$

$3!' = 108$

$4!' = 27648$

$5!' = 86.4$ million

$6!' = 4.0310784$ trillion

$7!' = 3.32 \times 10^{18}$

$8!' = 5.57 \times 10^{25}$

$9!' = 2.16 \times 10^{34}$

$10!' = 2.16 \times 10^{44}$

$11!' = 6.16 \times 10^{55}$

$12!' = 5.49 \times 10^{68}$

$13!' = 1.66 \times 10^{83}$

Table 10.1: The explosiveness of the Self Power Factorial (SPF) function

Translating to human terms – the number of cells in our body would be between 6!' and 7!'. The number of atoms in our body would be between 8!' and 9!' - and the number of atoms in the entire biosphere would be in the order of 10!'. The total number of atoms in the visible universe (estimated at around 10^{79}) is not even sufficient to take this SPF function into the teens!

A related series that is not quite as explosive - but relevant to this discussion - is the Self Power series: 1^1, 2^2, 3^3, 4^4 or N^N in general. The rationale for the naming – especially the 'Self' piece of it – is that the Number N is taken to the power of its own 'Self'. Whether the great mathematicians that penned the terminology 'Power' to describe the number of times a number is multiplied by itself knew about this future use, there is a key insight (albeit coincidental) to be had from this choice of nomenclature and is worthy of another small detour. The mathematics behind the Self Power function is developed further in Appendix 2.

The role of Intelligence in the Universe

If there were no intelligence in this Universe – the mechanics of the stars, planets, and entire Galaxy clusters would still go on like clockwork. Yet with no observer now, or in the future – it would be an empty existence – and all that would be left at the end is ever colder cinder blocks strewn about an ever expanding space. In the final reckoning, there would be nothing to distinguish such a burnt out eventuality from zillions of other universal dead ends – and would be discarded in the scrap heap of unfulfilled potential.

If there is an Intelligence that can establish itself at the Universal level using Yogic Universality – the whole mechanics for existence changes. A defining capability of Intelligence is that it can put 'planning' and 'design' into the future – and might even succeed in reversing the coldness and darkness that seem to be inexorably creeping up on us. In the event that our Universe is capable of spawning other progeny Universes – such a Universal Intelligence might be able to bundle some clues into the mix for a progeny intelligence to find. All the patterns we have noted, and many more that we are yet to discover – leads us to believe that the drive to better itself exists as a primal force within our universe – and we, humans, are an integral part of it. This force manifests itself in the form of Intelligence that is forward looking and scalable to take on the largest projects imaginable – even ones involving Universal preservation and potentially begetting highly capable progeny Universes.

In between the burnt out cinder blocks and the Universal Yogic Intelligence is the more likely possibility that lots of Intelligences appear but never realize their potential. Over time they either fade away through internal strife or ecological mismanagement – or just are not able to escape the laws of physics when it comes time for solar burnout. A subsequent Intelligence that does achieve Yogic Universality may be able to work through some of the fossils and artifacts – and file them away as specific learning as to things that could go wrong with a developing intelligence. If no future Yogic Universality emerges, then the end is not too far different from the cold and lonely cinder blocks that we discussed in the first scenario.

The core philosophy for an emerging Intelligence that aspires to Yogic Universality must have four essential characteristics. One is – it must be scalable across all the dimensions we know of (and maybe a few more). Second – it must be capable of finding common ground with other Intelligences it comes across – which we have referred to as Yogic capability. The third characteristic for any Intelligence that would aspire for the stars is a high level of internal organization and efficiency – which does not break when scaled or when in Yogic Union with another Intelligence or Entity. And the fourth key characteristic will be the need to keep alive an innate curiosity and a desire to better itself. As expected, these four characteristics are borrowed directly from the four segments of the Mahashrama Cycle. Thus, another way to define a forward looking Intelligence that has the potential to scale to galactic or universal levels is that it has really internalized the concept and practice of Mahashrama. As we go about discerning the internal design of the Supra Human Intelligence - these characteristics will continue to be critical design guidelines. Let us call a civilization that is thus endowed to undertake the transcendence of their civilization into a higher state of existence as being 'Self Powered'. With this small detour, let us go back to the discussion of the numerical factors that make up the number 108.

Levels of Existence, and the Tree of Universality

We can map the four segments of the Mahashrama cycle onto the 2^2 factor in 108. Let us call this level of human organization as Level 2. The 1^1 factor maps down to the realization of the essential unity of the purpose of Intelligence in the Universe – and that Yogic Unity is possible between all forward looking intelligences at the Universal level. We will call this Level 1. The 0^0 factor has deep philosophical significance – and it relates to our ability to transcend our existence to entirely new levels. We will refer to this state of transcendence as Level 0. The 3x3x3 factor has a lot of significance as well, and relates to the physical reality we find ourselves in – including space, time and the relationships that we see around us. We will call this the Level 3. Let us refer to the composite multi-level structure as the Tree of Universality.

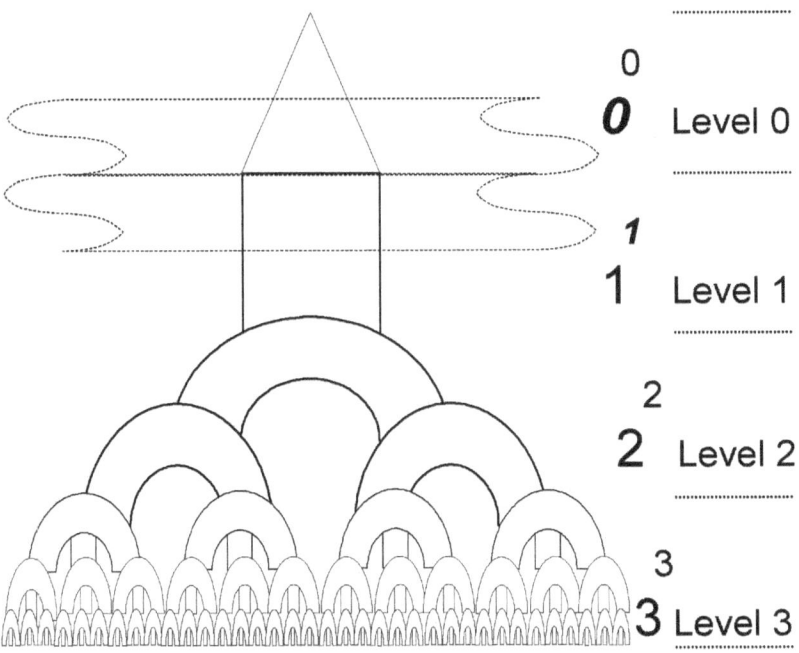

Fig 10.3: A graphical representation of the factorization of 108

Figure 10.3 illustrates how one might envision the full significance of the factors that make up the number 108 – i.e. $0^0 \times 1^1 \times 2^2 \times 3^3$. At each level we have the branching of the Tree of Universality corresponding to the factorization of the SPF series. Hence at the second level we have 4 branches, and at the third level we have 108. Level 0 is shown as a superposition of two lines – one that tapers down to zero, and one that is indefinitely large – but not necessarily infinite. This reflects the truism that $0^0 = N^0 = 1$ where N is any number, however large. Hence the factor seems to point to a spike embedded on an indefinitely large substrate. In Figure 10.3, the base term of $\boldsymbol{\mathit{0}}^0$ is indicated in italics and bold, as well as the power term of 1^1. We will develop this analogy further in the next section, so for now let us note that that Level 1 has an equivalent scaling characteristic as well, but this time for the power term.

Figure 10.3 is a visualization tool – but the reference to a large inverted tree descending from the heavens also comes from ancient Indian philosophy – and depicts the understanding of the essential nature of the Universe. In the Bhagavat Gita, Chapter 15, we have a reasonably detailed accounting of the functions of the various aspects of the Tree of Universality. In Appendix 2, we take a closer look at how Figure 10.3 relates to the Ashvattha or Peepul tree. Let us begin the discussion of the significance of the various levels - starting at Level 3.

Level 3 – Location Base

Level 3 appears to map down into the hard reality that we know as humans. Firstly, we visualize space in 3 dimensions. Second, the arrow of time inserts a trifurcation into everything that happens – either in the past, the present or in the future. Another way to look at the arrow of time is that for everything there is a time for 'creation', 'sustaining' and 'destruction'. Yet another way to look at the division of time is as a series of needs that express themselves, prompt action leading to their fulfillment, and the end state of fulfillment / gratification.

The third trifurcation, it would appear, relates to how we map our individual existence onto the Universe. One way to visualize our outlook is:

1. Our individual selves – the microcosm
2. Our environment that we can sense, and over which we may have some level of familiarity/control including the personal relationships we have with others – our relationship space.
3. The vast expanse of everything else in the universe that is hidden from our senses, or is very dimly expressed – the macrocosm

The use of this 3x3x3 branching can help us locate ourselves within a much larger existence – and gives us the physical coordinates or 'Location Base' using which we can give identity to the material part of our existence. This level of division is necessary if the Tree of Universality is to descend into communication (and possible union) with each and every sentient denizen of space-time. Our location in Level 3 is not only useful for sentient beings, but also other creatures that live in this material universe. Level 3 is relatively easy to map out with the tools of Science – and as such is the physical domain for our individual intelligence to make sense of, before embarking on an understanding of the levels above.

For a region in Level 3 space where a Yogic domain has already been established, the Tree of Universality can be visualized as the means of connecting the individual to the Yogic whole. The fact that the SPF (Self Power Factorial) as a function can expand very fast in just a few levels (Table 10.1) means that the individual need not feel too far removed from the Yogic commonality of which he or she is part. Building up to the level of the Supra, we should be able to use the internal branching suggested by the SPF to be able to quickly and effectively connect all the constituents of the Supra into an integral whole. So far we have discussed just a few levels of the SPF and how it might relate to organization within the Supra. The concepts can be extended to the fourth, fifth and sixth levels to accommodate a large population and distributed sensors and actuation points. Even at a full Supra

level of organization (comparable to the earth's biomass) – the seventh level is probably more than adequate to encompass every single sub-entity that is capable of independent thought and action.

Level 2 – Mahashrama

Rising above the day to day travails of our material existence, humans have the potential to create a reality that harnesses the power of both the competitive and the cooperative domains of human effort. This helps us organize society for both safety and prosperity, and over many cycles gives us the ability to consistently improve, or even transcend our existence. Most of the ideas in this book have been presented in a Mahashrama context. Let us summarize it here by noting that the 2x2 segment divisions, working together, sits on top of the mechanistic Level 3, and working in harmony can become our path to Yogic Universality (Level 1) and even Transcendence (Level 0). Let us take Figure 10.3 and name the levels to indicate the meanings that we have ascribed to each of these levels – which gets us to Figure 10.4.

Level 0 – A Paradox wrapped up in an Enigma

At first glance the term 0^0 tends to suggest that our whole existence is out of nothing and is intended to vanish into nothing. A slight bit of mathematical analysis – using the two zeroes as independent variables and observing how small changes perturb the value of the term – leads us to a big surprise. It turns out that the power term is VERY sensitive to the slightest perturbations! The value goes directly to zero or infinity depending on whether the slightest perturbation is positive or negative. Hence, there can be no doubt that this power number is precisely '0' and nothing else – as if the whole of the universe is balanced on a razor's edge. In the diagram in Figure 10.4 this represents the tip of the triangle that we see at the top and center.

If we do the same analysis to the base '0' we are in for a surprise as well! What we find is that the base '0' could be changed to any finite number, large or small – positive or negative, and the term x^0 yields the very same result! Thus:

Grihastha Vanaprastha
Brahmacharya Sanyas

$$1000^0 = 10^0 = 1^0 = 0^0 = (-1)^0 = (-10)^0 = (-1000)^0 = 1$$

The nature of Transcendence is such that incremental efforts can bring about massive changes in our outlook and how we live our lives. In the other extreme, however much we change our circumstances, certain things always remain the same – i.e. they transcend the limits of our very existence. In a mathematical way, the term 0^0 conveys both images of Transcendence. Because of these characteristics, the Level 0 of our existence is often very hard to understand – as it seems to present us with a set of contradictory messages, almost like a puzzle that we have to make sense out of.

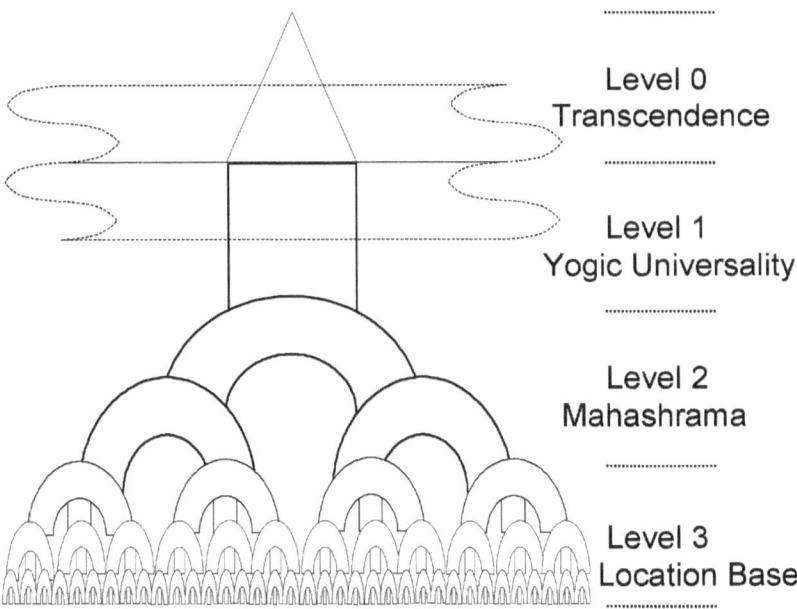

Level 0
Transcendence

Level 1
Yogic Universality

Level 2
Mahashrama

Level 3
Location Base

Figure 10.4: Tree of Universality – Levels 0–3 interpretation in human terms.

The significance of Level 0 appears to be that we have reached this level of consciousness where we can visualize our existence in more than one way. The first is the vanishingly small, a precise and mathematical zero – that teaches us humility in the face of

the magnitude of the Universe and what might lie beyond. The second, as we have observed, is an indefinitely large (maybe even infinite) scaling that we all appear to be part of. And most unusually, it does not seem to depend on how small or big we are as an individual or a civilization – in some ways we are still very small, and yet have the capability to scale upwards. Although the indefinite scaling property is shared with Level 1, the vanishingly small mapping is only possible in Level 0. It is for this reason that Level 0 is the realm of transcendence – where we try and fit our own meager voice into the wonderful symphony that defines the whole of existence, without feeling insufficient or inadequate against this vast cosmic backdrop.

Level 1 – Yogic Universality

In Figure 10.4 the base part of level 0 is indicated by the indefinitely large dashed line. This trait is shared with Level 1 – this time for the power term – which again could be indefinitely large or small, i.e:

$$1^{1000} = 1^{10} = 1^1 = 1^0 = 1^{-10} = 1^{-1000}$$

This lends credence to the belief that with a high degree of spirituality, Yogic Universality is indefinitely scalable to all intelligences that are Mahashrama capable, and the concept holds whether the domains are large or small.

In Figure 10.3 and 10.4, the two unusual quantities, the power term in 1^1 and the base term in 0^0, which share the same trait of indefinite scaling – are both represented by dashed lines. This pairing is so remarkable that it brings to mind a similar construct from cellular biology. The cell membranes are actually made up of dual layers of lipid molecules with well modulated porosity – and this dual walled structure is considered the key development that separates the 'living' organisms (like bacteria) from the non-living (like viruses). In the case of the human Tree of Universality, this dual scalable layering has very deep meaning as well. The first layer, the power term of 1^1 is the concept that Yogic Universality applies equally to all Intelligences. The second layer, the base term of 0^0 implies that all Intelligences have the potential to transcend themselves into a higher level of existence. We can imagine the two scalable layers as two separate dimensions – with each combination between the two layers capable of supporting its own tree of Universality. Envisioned in this way, the tree of Universality can turn into the forest of Universality, with each human (or machine intelligence) given the right to claim its own unique access to Universality – if they should so desire.

Bhagavat Gita Imagery and the relevance to the Tree of Universality

Illustrating from the Bhagavat Gita, the indicated path to achieve union with Krishna (the Supreme Being) is to achieve a high degree of spirituality, and then to cut off our reliance on the tree-trunk of Universality that we have just climbed. The interface of the two indefinitely scalable layers (i.e. the interface between Level 0 and Level1) is likely the break where one can abstract out a specific instantiation of the Tree of Universality. Let us refer to this interface as the 'veil of abstraction'. According to the (Bhagavat) Gita, Spiritual Transcendence can put us in a position to visualize existence in its true form, integrated over that which exists now, existed in the past, or would transpire in the future. Developing from the Gita message, as humans strive towards Divinity, it might no longer be necessary for the sanyasi to permanently 'cut off' their personal connection to the Tree of Universality. Yes, the sanyasi will not spend as much time connected to the nerve center of the emerging man-machine civilization as the vanaprasthi. Yet, there are critical areas that the sanyasi will need to continue to contribute to humanity – namely the preservation of the Gems of wisdom, the creation of a common human cultural genome, and the laying out of the vision of success for humanity in our quest for Divinity.

In Indian religious philosophy, the veil of abstraction is also called Maya – and the part of the Tree of Universality below (i.e. our physical reality) is often referred to as an illusion. What we have tried to do in this Chapter is to show that it is not necessary for our physical reality to become a 'throw-away' i.e. that it is possible to connect our mundane existence to the root of universal spirituality, and to be able to move back and forth as the situation demands.

Even in the Gita, it is indicated that the spirit of Krishna would sometimes come to the real world as an Avatar (an earthly form of a Divine entity) – which indicates that the 'veil' is designed to be bridged from time to time. Yet, bridging this veil appears to be a very complex task, which we will discuss further in Appendix 2.

Universal Success and Transcendence

For a moment let us envision ourselves at the very peak of universal accomplishment, and say that after several billion years of effort - mankind's message of Yogic Universality has been subscribed to by every single civilization within the entire Universe (as we know it). For the sake of argument, let us also say that the secret to having our Universe exist in a viable form for perpetuity has been unlocked, and the various constituent civilizations are working like a single organism to bring that vision to reality. We look back at our collective accomplishments and see a universal citizenship teeming in diversity and creativity, yet able to synchronize our resources to the tremendous tasks at hand. Are we at the end of the road, and can we consider ourselves 'Successful' as a Universe-wide Intelligence?

The Level 0 (Transcendence) of our human psyche, and presumably of a Universe level entity in a few billion years – will always wonder: is there more to existence beyond what we can sense or even imagine? This being the zone of Transcendence, we can imagine our existence vanishing into a point of nothingness, yet become aware of a relentless torrent of possibilities that may be out there – somehow eluding our grasp to quantify and subject to the rational laws of science. If we were but one Universe on the path to a peaceful and constructive end-game, might there be a million others in the throes of destructive internal conflict, and never able to achieve their full potential? Or could there be others with such tremendous accomplishments that they could have found a way of casting their positive influence onto 'our' Universe…

To some folks, opening up Level 0 might look like bringing in an unwarranted influence into a very explainable universal existence. Having gone into all this effort to determine that irrespective of our individual spiritual and religious beliefs we can all work together towards a scalable Yogic Universality – why open it up to random and unquantifiable influences from Level 0? Would not all the focused energy from the Mahashrama Cycle be for zilch if everyone could take their direction from Level 0 and come up with their own interpretations of what the 'Spirits' are trying to tell them?

In some ways, this denial of Level 0 is similar to the beliefs of the 'Flat Earth' proponents in the middle ages. In their beliefs, if ships sailed far enough away they would come to the edge of the earth and drop off a tremendous cliff! Who in their right mind would want to go anywhere near this 'point of no return'? The intrepid explorers that went further and further out and demonstrated that there was no such physical limitation – could only do that if they had an inner system of beliefs that went beyond the current cultural mindset. To have a good understanding of who we are – especially as an universal intelligence, it will be important to have some ideas as to the boundaries that define our realm of existence. As we pursue a combined man–machine future for the human civilization, this is also one area that the human part of our progeny are expected to excel at, as we will discuss shortly.

The secret to being successful in our contemplation of Level 0 (or Transcendence) is that we have some level of mastery of Level 1 – or the concept of Yogic Universality. This, in turn is based on a keen mastery of the concepts of the Mahashrama Cycle (or Level 2). To an individual or entity that is still trying to cope with Level 3 (stimulus-response from our current location in space-time, or material existence) issues – contemplating on Level 0 is like a caveman trying to exposit on the notion of whether the world is round or flat. There would be a severe lack of observed knowledge upon which the caveman (or any Level 3 bound individual) would be basing their Level 0 extrapolations – and hence would not carry a lot of credibility.

Of course, the aforementioned caveman does have an opinion, but how much value would today's explorers of our solar system put on such an opinion? To the caveman, however, his notions of how the Universe works does have value – hence our concept of Level 0 ends up being very personal and unique to what we consider as knowable in this Universe. Thus, we can conclude that although for ourselves our Level 0 perceptions have value – it only has value to others if they can relate to it as a natural trajectory from their own Level 2 and Level 1 experiences. With sustained effort, we will find - what is considered unknowable today, becomes knowable tomorrow; and the boundary keeps shifting out over time.

Human spiritual behavior throughout the ages seems to suggest that some acknowledgement of 'Transcendence' is required for us to succeed as a Global, Galactic and ultimately Universal intelligence. This seems to allow a peaceful population to be intensely competitive when it comes to the control of assets here on planet earth– yet highly cooperative when it comes to spiritual matters. Our greater goal now becomes to make it possible for all forward looking Intelligences to reach Yogic Universality, connecting to the tree of Universality at Level 1 – and subsequently seek to define an even higher level of existence. This is a recursive and iterative process, as the vision of transcendence today can become the Yogic reality of tomorrow, and the cycle goes on increasing in scope until we have reached the scale of the Universe.

Transcendence and Meditation

The expression of Transcendence on our rather limited existence is probably best realized through meditation. There are many forms of practice that meditation takes, but a common end goal that is attainable is a perspective of how our relatively puny existence fits within the greater scheme of things. Trending our ego down to a level close to zero during our meditation really helps us gain an unbiased perspective of the relationships that surround and nurture us in our quest for greatness.

One beautiful thing about Transcendence is that we can imagine a Universe that is filled with incredible accomplishments and is sustained by a high state of happiness (or Ananda) by all of its participants. Just opening ourselves up to such glorious possibilities through meditation fills us with a spiritual joy - that can be quite as intense as any physical pleasure we could give ourselves, and much longer lasting.

We have already observed that Transcendence is individual and unique to the person or entity practicing it. The wonders of scalability that have worked so well for us through Level 1 (Yogic Universality) – are no longer as relevant in Level 0. The economies of scale from collective thinking diminish after Level 1 – and a jump to Level 0 is always very individual. It may even be downright detrimental to get too much of our transcendental bearings through others – especially if they were derived from level 1-3 realities that are far removed from the ones we see around us. The historical and demographical realities that have catalyzed the great works of transcendental insight like the Bible, the Koran and the Bhagavat Gita also need to be interpreted in light of the expanded knowledge and wisdom that are available to us today.

Transcendence and Machine Supra Cells (MSCs)

As we have noted in Chapter 9 (Figure 9.2) there are aspects of the future progress of civilization that humans will be good at – and others that our machine counterparts, the MSCs (Machine Supra Cells), will be good at. The contemplation of Level 0 Transcendence is likely to demand skills towards the top of the range pictured, and is therefore likely to remain the relative forte of humans for a long time. A similar chart, Figure 10.5, may now be drawn up showing how our two sets of progeny's natural talents lend themselves towards challenges at the various levels of Mahashrama.

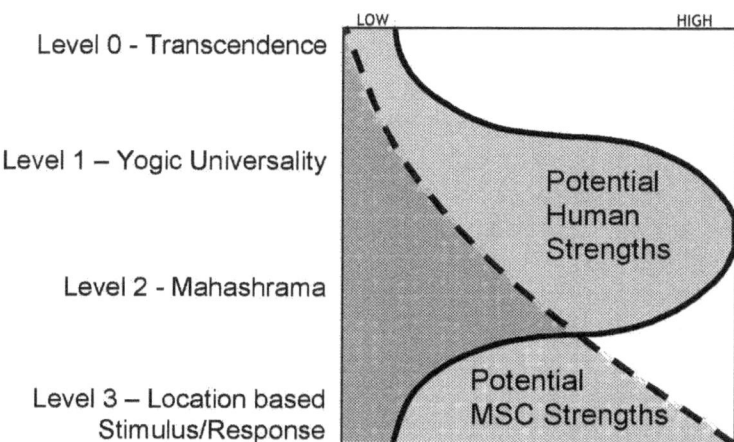

Level 0 - Transcendence

Level 1 – Yogic Universality

Level 2 - Mahashrama

Level 3 – Location based Stimulus/Response

Fig 10.5: Projected Human vs. MSC strengths – by levels on the Tree of Universality

The interesting difference here is that as we transition from Level 1 to Level 0, – we can expect Human capabilities to drop off significantly. Even then, the Human Psyche is significantly better adapted to deal with the abstract concepts and apparently contradictory messages of Level 0 than the MSC's. Human aptitude will likely peak as we bridge between Levels 2 and 1 – and Transcendence will likely continue to be a challenge for most of us with our current Cultural Genetic mindset. Of course, it is possible that we will run into civilizations that are much better at contemplative and meditation skills than we are. Not awaiting that eventuality, we must also make an effort to preserve the transcendental diversity that we already have within our own Human Civilization – for example the monks and mystics that make their homes high up in the Himalayas. Without significant effort, this is an endangered lifestyle in the world today (2006). Our human civilizational diversity today labors under the shadow of a tremendous cultural wave, modeled after the materialistic Western values, which appears poised to take over our entire planet.

Going forward, when we have man and machine working closely together to create our future reality, the first stage of specialization between Humans and MSCs would likely start out as follows. Level 3's Location based domain will almost certainly be the first domain where MSCs will make their presence felt. When it comes to Level 2 – or harnessing the developmental energy of the Mahashrama Cycle – MSCs and Humans will share complementary talents – with significant contributions of MSCs in both the Grihastha and Vanaprastha segments, and lesser contributions in the Brahmacharya and Sanyas segments. The rationale, here too, is that the opportunities in the top two segments are much easier to quantify and convert into actionable tasks than the proliferation of ideas and skills that characterize the Brahmacharya segment; or the values of condensation, simplification and Cultural DNA encoding that marks the Sanyas segment.

By the time we collectively reach Level 1 and really grapple with Yogic Universality – one might think that MSCs with their muted egos will perform better than Humans. But the more likely outcome is – the level of abstract thinking that is required to really harmonize several diverse individuals, cultures or civilizations will be very hard for MSCs to come by (at least for a while). In a way the transitive nature (limited life span) of human existence also comes to our assistance. In the second half of our natural life, when we are well into our Vanaprastha phase of existence – our egos become more receptive to a Level 1 transformation, which should allow us humans to work very well towards Yogic Universality.

Level 1 Transformation (Yogic Universality)

We have described the beginning of the Vanaprastha segment as the realization that we have discharged our Genetic imperative (viable, well situated offspring), and that we are now willing and able to widen our circle of caring (and action) outside of our own family and onto the world at large. We have discussed how the circle of caring may be as narrow as one's own community initially, but could grow to include the entire human race, the entire biodiversity on Earth, or even larger spheres of influence. Let us take a moment to identify the markers that indicate that we have

been able to grow our perspective enough to transition mentally to the realities of Level 1 or Yogic Universality – which we will refer to as a Level 1 transformation.

Yogic Universality begins with the practice of the Yogic Empathy skills described in Chapter 4. The understanding of who we are and what our aspirations for the future are – is the first step towards Yogic Empathy. When two entities with a solid understanding of who they are (and their mutual aspirations) work together to create a combined reality that is greater than the sum of the parts, and are able to work almost as part of the same organism – we move from Yogic Empathy towards Yogic Union. The Yogic Empathy skills are much the same whether they are practiced between individuals, or ethnic entities, or even within our own selves. Yes, the same skills can be practiced and perfected between various parts of our own existence – say between our body and mind, or even between our Scientific and Spiritual mindsets. A person at peace with their own internal priorities is much better able to extend the skill to others – and perceive the whole of existence as a series of Yogic Unions that ultimately can lead to Yogic Universality.

Yogic Universality, Universal Intelligence and Life

The person or entity at the brink of a Level 1 transformation has to not only be able to scale upwards to identify with the concept of a Universal Intelligence – but must also wrestle with how far down the intelligence scale they'd wish to go. Beyond humans, are household pets parts of an ecosystem we can create an yogic bonding with? Can we do the same with animals in the wild or even with animals meant for human consumption? How far do we go - down to insects as the orthodox Jains believed – and keep our mouths and noses covered lest we accidentally breathe in and kill an insect? Or does the circle of caring go all the way down to the unit of life as we know it – the cell?

A distinction will need to be made between the units of life that we consider worthy of our care, and ones that we can consider sophisticated enough to be part of the Supra Cellular network. We have, rather arbitrarily, assumed that a Machine Supra Cell has

to have about the same level of intelligence as a human, but why not a chimpanzee or a mouse? A much more robust definition is of an intelligence or entity that can understand the concept of the Mahashrama cycle, and modulate their activities according to the segment that they are in. If the entity or person has the capability to achieve a successful Level 1 transformation – they are the units of which a Universal Intelligence can be constructed.

That will leave, of course, a plethora of living creatures that live their life primarily as a function of their genetic programming – and not through learned or acquired cultural programming. The vast amount of the biodiversity on earth will fall under this category – living primarily in the left half of the Mahashrama Cycle. Somewhere below the level of the individual cell – we enter the realm of in-animate objects – and the level of organization drops below our circle of caring or biological kinship. The person contemplating a Level 1 transformation will need to make an individual determination of what constitutes good stewardship of the biodiversity that we see around us. For those inclined to be too exclusive in the determination of the thresholds for intelligence and caring – please consider that one day we will need to present ourselves in front of intelligences (quite possibly of our own making) that are so superior to our own that we ourselves would look like individual bacteria in their relative perspective.

Level 0 Transformation (Transcendence)

When it comes to man and machine working together on our next few levels of encapsulation – we have noted that it will be the human element in us that is likely to focus more on the Level 0 transcendental elements of our existence. With transcendence, the vanishingly small can beget the irresistibly large – and normal science, which deals with the interrelationships between measurable quantities, begins to lose its applicability. Just as the genetic mutations that over billions of years yielded organisms as complex as ourselves, were themselves vanishingly small – similarly the ideas and discoveries that cause a change in our Cultural DNA come from vanishingly small and sometimes quite unexpected sources. You may have heard the story of Friedrich August Kekule who in 1865 deduced the internal construct of the

benzene molecule (chemical symbol C_6H_6). Chemists of the age were puzzling over how such a stable compound could be created using the rather low ratio of Hydrogen to Carbon atoms. One day, as Kekule was dozing off – he imagined the carbon chains as snakes that were in rapid motion. At one point one of the snakes caught a hold of its own tail and started spinning rapidly. Kekule immediately realized that the chemical structure he was looking for was actually a ring with six Carbon atoms connected together – each attached to a Hydrogen atom.

This rare piece of insight coming when the inventor least expected it – is in many ways representative of a Level 0 transformation. It is as if we cast our net wide for the most ephemeral of ideas – but retain the capability to latch on to the most promising ones, almost instinctively, when we find them. To test these ephemeral concepts, we quickly subject these ideas to the realm of science and measurements to determine their domains of applicability. Even if we find a partial applicability, we file them away for future use in case one day a domain evolves where the concepts are a natural fit. For example, credit is often given to the Italian mathematician Fibonacci from the early thirteenth century for inventing the math behind compound interest – even if the concepts were not to find widespread usage in his day.

Just as the Level 1 transformation is a culmination of the Vanaprastha stage of our existence, a Level 0 transformation is a culmination of a successful Sanyas stage. If we look at the last part of Sanyas in Mahashrama Cycle, the economic footprint can be seen trending to zero. A successful sanyasi is adept at juggling ideas and concepts that may not have any economic significance for a very long time. Through transcendence the Sanyasi can relate to different realities, some of which are about how humans might be able to cope with the most pressing realities of the day, and others to the realities of a time many centuries hence. The time frame does not need to be only forward looking. A sanyasi that spends his or her last working years condensing down to basics what it means to be the last practitioner(s) of a disappearing lifestyle – is dealing with vanishing realities in the past tense.

Many mystical powers have been attributed to sanyasis who have truly mastered the Level 0 transformation – giving them capabilities that seem to defy the laws of physics. In a quantum universe – violating laws of classical physics may be looked upon as bringing forward a highly improbable set of eventualities. Even without the help of magic or miracles, the bringing forward of favorable outcomes that would otherwise have a vanishingly low probability of occurring – is the realm of the accomplished sanyasi who has successfully mastered the Level 0 transformation.

Going back to the roots of Greek Philosophy with Socrates and especially Plato – there is a belief about an Ideal Plane of existence inhabited by perfect forms. Our body and the material world around us - are assumed to be a deformed instantiations of the idealized forms, with only our souls being able to travel back and forth between the idealized world and the material universe that we perceive around us. This dichotomy of body and soul is at the root of many religions, including Christianity. If we equate the Perfect Forms of Plato with the Ideas and Concepts space that we humans can discover after a successful Level 0 transformation - there no longer is a need to forever separate out the gross and imperfect from the spiritual and the sublime. They are but two aspects of the same Tree of Universality.

The grossness of reality we see around us, as well as the pain and suffering that seems to be associated with almost all aspects of existence (even something as noble as child-birth) - is the reality of a Level 3 and most of a Level 2 existence. In a competitive world, building skills takes pain during Brahmacharya, putting in sustained effort during Grihastha takes pain (including the aforementioned childbirth) as well as dealing with a painful past and present and trying to coax humanity to better itself (Vanaprastha)! Yes, there is much truth in the saying "No Pain, no Gain". A lot of these painful experiences leave their marks upon us; small deviations from the 'Ideal' genotype that would represent us - that together add up to create the 'Grossness' that we associate with reality. In fact, on the rare occasions that we find something truly beautiful and pristine – we tend to invoke divine metaphors and similes – e.g. like an Angel; a thing of beauty is a joy forever; divine beauty, etc.

It is only during the Sanyas stage of our existence, especially after a Level 0 transformation, and having come to terms with the transitivity of life, can we free our minds sufficiently to start thinking in terms of pure ideas and idealized forms - without having to bend them to suit our ego or adapt them to our station in life. In this space, we will find concepts so powerful that they would outlast the universe itself, or normative ideas that would enable diverse intelligences and civilizations to work together and trust each other explicitly and implicitly! Like the long poles in a tent appear to be slightly taller than the tent itself, it is these grand ideas that could form the core structure of a universe wide intelligence – and as such must be held to have meaning even at the very ends of our own existence.

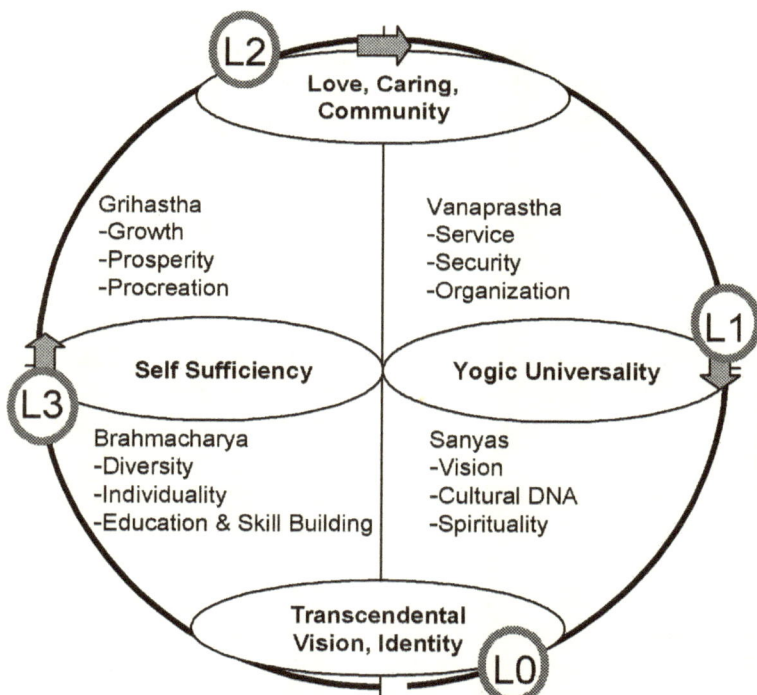

Figure 10.6: Transformation Levels 0-3, and the associated Mahashrama transitions

Level 2 Transformation (Mahashrama)

Our Level 1 and Level 0 transformations, as we have noted, are unique in that they connect us to the infinitely scalable and the vanishingly small. Figure 10.6 maps all the levels of transformation onto the corresponding segments of the Mahashrama cycle. As expected, the Level 0 transformation happens when we have gained a degree of proficiency in the Sanyas segment, and Level 1 when we have mastered the Vanaprastha Yogic principles. In addition, there are two other major transformations of our outlook that happen at the end of the Grihastha and Brahmacharya stages – to reflect our mastery of the key learnings from these segments, and a preparedness to move on to the next level.

The Level 2 transformation (L2 in Figure 10.6) comes towards the end of our Grihastha stage – and reflects the inner realization that there is more to life than our individual circumstance; i.e. our selves, our families and our close circle of friends. We have seen earlier that the Vanaprastha transition is particularly difficult – and many people never get out of the reassuring comfort of cycling through personal need fulfillment and asset accumulation. Yet, to those amongst us who do see meaning beyond our personal fulfillment cycles - we have handed the flag of Vanaprastha service leadership. These are the folks who get to carry the message of love, caring and community out in ever widening circles – and make a determined drive to pull humanity towards full Yogic Universality (Level 1 transformation).

Yes, the Level 2 transformation is the one that starts us towards future success with Level 1 and even Level 0 transformations. Hence, we can look at the Level 2 transformation as the gateway to the uniquely human part of the Mahashrama Cycle. We have noted already that animals typically do not get a chance to participate in much beyond their genetic imperative, and the natural selection process really is not very conducive to altruistic long-term thinking. Equipped with an outlook change corresponding to the Level 2 transformation – we humans now have the ability to apply not only our talents, but also our considerable social standing and economic resources to creating a better future for all of humanity.

Mahashrama success, and the Tree of Universality

Looking at Figure 10.6, the astute reader has likely already figured out that there is a correlation between the four stages of the Mahashrama Cycle, and the four levels of the tree of Universality that we have defined in Figure 10.4. By this correlation, L2 also coincides with our proficiency in the second level of the tree of Universality – which corresponds to the Mahashrama Cycle itself. One might ask - how is it possible to be proficient in the full cycle if we still have two more segments to go? If we define L2 as the gateway to the right half of the Mahashrama Cycle – and a deep internal realization that there is more to life than our individual selves and our Genetic Imperative – we have indeed internalized the underlying principle of the Mahashrama cycle. We have made the 'leap of faith' to move beyond our individual circumstances – and are now sure that there is no cliff that we will fall off as we embark upon our journey to circumnavigate Mahashrama.

Level 3 Transformation (Location Based Self-Sufficiency)

We have noted earlier in this Chapter that the Level 3 of our tree of Universality is the realm of all that is real – in space, in time, and in the relationships we find ourselves in. Mastery of Level 3 reflects our ability to succeed in knowing the reality (Brahma) around us, and in having a certain degree of proficiency that allows us to take care of ourselves within this existing reality. In Figure 10.6, the elevator from Brahmacharya to Grihastha is called 'Self Sufficiency'. This is the gate that allows us to leave behind the net dependency we have on others during Brahmacharya, and become a net contributor to others (including, potentially, a new family unit) as we proceed further with Grihastha. For a human being getting ready to circumnavigate Mahashrama, this is the first gate that we need to pass. It is not expected that the graduating brahmachari can fully internalize the concepts of a Level 2 transformation at this stage – because they have not yet had the opportunity to grow their individual genetic and economic footprint. Yet, even a cerebral understanding of the journey ahead goes a long way to preparing the graduating brahmachari of the fullness of the potential that lies ahead. Thus equipped, they would be able to recognize the need fulfillment behaviors of all four Mahashrama segments (as summarized in Chapter 4). This helps them balance

their need for personal goal fulfillment with consideration of the needs for others - and thus prepares them to be exemplary citizens.

We have noted earlier that criminals and other disruptive individuals who do not respect the core needs of humans across the Mahashrama segments, may not be given the honor of a Level 3 Transformation and the privileges of citizenship that come with it. Thus, in more ways than one, Level 3 is a rite of passage – into a physical and intellectual maturity that allows us to embark upon the most productive and personally rewarding segment of Mahashrama – namely Grihastha.

The Yogic Planet

Let us change gears and take a look at spirituality from the perspective of human ethnicities and cultures – and the spiritual worldviews that we collectively subscribe to. In Chapter 6, Figure 6.8 we have mapped out the religions of the world into 2x2 segments, depending on whether they subscribe to a single unifying Divinity, and whether they believe that Yogic Universality is possible for our human civilization. Let us take this opportunity to envision how the Tree of Universality might look like under the different worldviews, as represented in Figure 10.7.

Yogic Universality Possible?

Pre-existing, Organizing, Divinity?	No	Yes
Yes	Forest of trees **Grihastha**	World's Biosphere (Earth as trunk/monolith) **Vanaprastha**
No	Fertile Grassland **Brahmacharya**	Entire Universe (Integrated over Time, Space and Probability) **Sanyas**

Figure 10.7: Tree of Universality perceived through different worldviews

We have noted in Chapter 4 that 'Spiritual Self-Determination' is a core Sanyas need – and our Spiritual worldview is a direct result of our having satisfied that need. It is the influential sanyasis (which includes prophets and other spiritual leaders) that have typically imbued each culture or religion with their worldview. It is possible that a person's spiritual worldview changes as they move from segment to segment – but the society, taken as a whole, typically subscribes to a worldview (or set of worldviews) that change very slowly. As we noted in Chapter 6, most of society is today split between the individualistic Brahmacharya worldview and the hierarchical Grihastha worldview.

The Brahmacharya viewpoint of spirituality, where every man is an island, can ultimately lead to a level 0 transformation – but the tree of universality would look like individual strands connecting each person to the cosmic continuum. Hence, with the scaling in two dimensions applied at the top of Level 1 and the bottom of Level 0, this two dimensional space looks like an expanse of grassland, with each thread of grass individually embedded in the substrate. In the Grihastha segment we have multiple trees of Universality, each depicting a set of beliefs (religions, sects, denominations) as to how we can connect to Divinity. Hence, the terrain now looks

like a forest with multiple trees with deep roots each supporting a large number of adherents.

In the Vanaprastha segment, we seek to combine together all the human intelligence that we see around us – and the universal tree trunk becomes a vast monolith. One can think of the Vanaprastha model as the biosphere girdling all of earth, and Mother Earth herself becoming the conduit to Universality. Let us take a slight detour and investigate the analogy a bit further.

In many ways, this is similar to the Gaia theory[2] – where the world's Biosphere is looked upon as one interconnected self-regulating organism. We have already drawn the parallel that shows the Earth's Biosphere is about equivalent in complexity to the Supra Entity. To take it a step further, the Santiago theory suggests that a 'brain' is not necessary for a 'mind' to exist, and that the vast interconnected link of living protoplasm already gives our Biosphere a 'mind' and associated cognitive ability. We do not have to go quite as far as to include bacteria as part of the Universal Mind – but the concept is illustrative of the essential unity of all intelligence. We have already defined a 'forward looking' intelligence as one that can move purposefully past their L2 transformation – so, for now, this sets a limit for how inclusive we need to be to be considered intelligent. Not surprisingly, today this draws a neat circle around humanity – but the dynamics would change if we were to discover another intelligent species.

With the Sanyas worldview, we are now in the realm of the intelligences that we don't yet know, ones that might exist already in a form that we cannot recognize, or even be of our own making (in the future). Here we can consider the possibility of the earth's Biosphere already having a mind of its own, or even individual species through their common genetic thread. We can attribute a potential 'mind' to thinking patterns in the plasma of the Sun – and of the other stars and black holes within this Universe. We can envision the wisdom of the universe beings stored in the way of ultra-low frequency electromagnetic and gravitational waves that permeate the universe. If the Vanaprastha view of the Tree of Universality is the Yogic Planet – the Sanyas approach is to set out a Vision by which all of the Universe can be (or already is) permeated with Purpose and Intelligence. This Purpose is not

static, but continually seeking to improve itself, and transcend into realms that were previously considered unachievable. The vision of the planet girdling biosphere as the tree of Universality (derived from the Vanaprastha segment) now becomes a universe wide entity, with no in and no out, and with nothing excluded. In short, the Tree of Universality now becomes - the Universe!

Spiritual Growth for Society

The facility of mind that is needed to envision the Tree of Universality in the Sanyas worldview is indeed tremendous, and gets especially difficult as we cannot use the familiar measures of 'Up' or 'Down' to check if we are moving towards the core of spirituality, or away from it. For example, if we are too inclusive of what we consider intelligence – we might even be tempted to give up our existence so that the bacteria that will feast on our body can have a good time! A keen understanding of what constitutes the 'betterment of the Universe' is called for – which can only be based on an equally keen understanding of how the Universe already works. It is for this reason that the Sanyas worldview is especially challenging, and takes deep insights as to how our universe operates to be successful.

The Vanaprastha worldview, having defined Intelligence in relatively human terms (including the capabilities of the MSC) – allows us to seek out the commonalities in a more familiar context. The fact that this worldview is friendly towards the presence of a pre-existing, organizing Divinity (or God) – is consistent with the need that a lot of humans have to believe in a higher power (or God). As we go about stretching our concept of existence - this is probably the worldview that most of us humans can aspire to. As noted earlier, success with the Vanaprastha outlook depends on the mastery of Yogic Empathy – which is itself a considerable stretch for us highly competitive humans. Ultimately, when we get to the Supra level of existence, it is possible that we will garner enough intelligence as to how the universe operates, and that we can collectively move our worldview over to Sanyas. Until then, the realm of the Sanyas worldview will likely remain the domain of the inspired thought leaders; and these will be the people who will paint the vision for how we humans can keep expanding our

vision of inclusiveness - to embrace more and more of our growing perspective on this universe.

An interesting parallel can be drawn from the study of Buddhism which draws its worldview primarily from the Sanyas segment. In many places, Buddhism has mingled with the existing local beliefs to create a pantheon of lesser deities that are worshipped as a stepping stone towards becoming a Buddha. It is reported that in Myanmar (Burma), the 11th century king Anawrahta designated 37 local spirits or 'Nats' as deities that were officially sanctioned part of the local Theravada Buddhist Pantheon. (Source: National Geographic, May 2006). This was done only after trying to ban their worship altogether – a move which was resisted by the population. In Tibet, China and Japan, Buddhism has also come to co-exist with a pantheon of demi-gods derived from the local culture and history. This phenomenon does not take anything away from Buddhism as a pure worldview, but does indicate that there is a basic human need that can only be met by having a God-like entity that we can look up to for guidance and protection. When humanity creates our own God-like entity (the Supra) – it is possible that we will finally overcome this psychological dependency.

As we have noted earlier, much of humanity today subscribes to the Grihastha worldview. Here we have likened the Tree(s) of Universality to be more like a forest, with many trees representing the various schools of spirituality, churches, sects and denominations that exist in the world today. Each tree in this forest can potentially connect thousands of humans to their spiritual roots, but a mechanism does not exist for them to join forces and become one. Divided in this way at the root of organization, they each serve a noble purpose. However, as currently constituted, very few will find it easy to extend the doctrine of their forefathers - and accept that there is a higher order of existence that unites all the various beliefs. If enough people subscribe to the unifying Vanaprastha worldview, even these venerable institutions will likely begin to change, and the great debates about the shared nature and purpose of humanity can start! We will see in Chapter 11 why such a process is critical for us - to create an identity for ourselves as the 'human civilization', and thereby take control of our own destiny.

Chapter 11
Conclusion

"Lastly, to be fully is to have the full delight of being. Being without delight of being, without an entire delight of itself and all things is something neutral or diminished; it is existence, but it is not fullness of being."
- Sri Aurobindo, The Life Divine

In this concluding chapter we will briefly touch upon the awesome potential that we find latent within ourselves and our environment. We hope to internalize, one more time, that humanity (and our progeny) can be part of the main act within this Universe. Based on this assumption, let us validate once again that every system of belief amongst humans today can be integrated into a greater context based on Mahashrama – even something as contentious as religion. As we press forward with the increasingly technological content as to how we define our lives, we will marvel once more at how our biological and technological progeny will be so amazingly complementary in their capabilities and aspirations – that together they can be an irresistible force as we chart a future that reaches for the stars.

With Yogic Universality as a goal, and planet Earth as our environment within which to germinate the next level of our existence – it would behoove us to look at some of the shorter term tasks as well. One of the first steps we could take as Humans is to set up a strong Vanaprastha Network that supports us all, and directs our major efforts of self improvement and long range planning. When all is said and done, we will establish that the future that we are setting out to attain (and even if it has been attained before by a parent civilization to Humans – who we would know as God) is so glorious that it is a source of eternal Spiritual Joy or Ananda. This is the 'full delight of being' that Sri Aurobindo refers to in the quote at the beginning of this chapter – and offers a relatively simple metric to track as we strive towards self-made (or inherited) divinity.

Planting of the Tree of Universality

Ancient Indian visionaries have often visualized the essential
nature of existence as a tree with many branches descending
from the heavens. If we take the Tree of Universality (Figure
10.3) that we have defined as our instantiation of the descending
tree envisioned by the ancients – it is important to determine that
this tree is well planted and has firm roots. If the Level 0 reality
(Transcendental ideas, beliefs) upon which we build this Universal
tree is itself smaller in scope than the Universe we operate within
– we might find ourselves planted in shifting sands. On the other
hand, if our core beliefs are scalable to a host of alternate realities
(some of which may not even exist in our Universe) – then the
Level 0 substrate that holds us can be expected to support
our Universal tree through the various stages of maturity and
fruitfulness.

As we look across humanity today, we find many threads of
Philosophy (and Religion) that have each survived and prospered
in a variety of different settings (time, geography). As we
attempted to integrate their core belief systems (Chapter 6) we
found that there is a structure and pattern that runs through them
that fits well in a greater Mahashrama context. Similarly, the core
concepts that will form the underpinning of our future existence
must morph and scale with our expanding understanding of
reality. We have already noted (Figure 10.2) that neither man nor
machine in today's cultural environment is very well suited for
Transcendence (Level 0). Is it sufficient to leave the building of
these skills to the elderly Vanaprastha contemplating the move
into Sanyas – or is this something so critical that we should teach
the essential skills to our children even if they do not get its full
use for 50+ years? In fact, the same question can be asked for
the Level 1 transformation skills (Yogic Universality).

The author, personally, was introduced to the practice of Yoga and
meditation at an early age, and discovered them to be very useful
during stressful times – whether as a teenager, as a newly married
couple or a parent of an (almost) teenager. The skills themselves
are surprisingly independent of any religion (including Hinduism
from which they are derived). From the standpoint of mental

development, these core concepts and their practice should be able to be practiced by individuals approaching their teenage years – and possibly much sooner under the guidance of a Yogic expert.

The planting of the tree of Universality starts with the sowing of these core concepts in the developing mind – so that over time the individuals will welcome (rather than fight) the transitions from Brahmacharya to Grihastha to Vanaprastha and then on to Sanyas. The realm of Level 0 (Transcendence) is in our minds, and the earlier we are exposed to it, the earlier we can start putting together the 'Long poles' that would define the structure of our individual and collective existence.

An early and widespread exposure to the Level 1 (Yogic Universality) and Level 0 (Transcendence) concepts would also stir a vigorous debate as to the nature of the long poles of our human existence, and even how we may relate to other potential intelligences within this universe. This normalization of the ideas around the meaning and purpose of our existence goes a long way towards defining our Cultural DNA as a race. Until most of humanity subscribes to one (or at most two) internally consistent threads of who we are and why we exist – mankind cannot be said to have developed their identity as a civilization. The case for two threads for certain critical areas may be needed – as we have discussed in Chapter 8. This is for the sake of genetic diversity, and should be comparable to the genetic diversity we have between our own pairs of DNA sequences as illustrated in Figure 8.5.

There is relevance to a unified human cultural DNA, potentially, to other nearby intelligent entities as well. Should there be an external intelligence watching over us (or just passing by) the normative process of determining our cultural DNA might seem like an interminable wait as our whole civilization grapples clumsily with their long poles. Yet, it can be argued that any civilization which has the inherent potential - deserves to complete the codification of their own identity and purpose; as a major step towards fulfilling their civilizational destiny. An external entity that

tries to help too soon would risk polluting the civilizational gene pool, and likely end up suppressing the existential gems that make the emerging civilization unique and vital.

We can look upon this process of normalizing a common identity and purpose as the planting of the tree of Universality in our collective human psyche. This is the coming of age of humans as a civilization – and will shape how we see ourselves as an earthly, solar, galactic or even universal intelligence.

Divide and Conquer?

Coming back to the question of when it is that an external intelligence should make their presence known to us – we have just defined a threshold of identity and purpose that would mark our maturity as an intelligent race. So why is this threshold important? Let's consider what would happen if a vastly superior intelligence were to make themselves accessible to us at this juncture of civilizational growth. In a Level 3 (Location Base) or even a first half of Level 2 (Mahashrama) existence – the first thing that would happen amongst us humans is a power struggle as to who would be the interpreter-spokesperson to begin a dialog with these highly capable beings. Once the channels have been consolidated (or multiple channels created), there would begin the process of aligning ourselves with the power centers – and the keeping for one self the benefits that came out of trade (or other exchanges) with the external intelligence. Conquerors throughout the ages have used the principle of 'Divide and Conquer' to their advantage – and it is likely that an advanced intelligence would prefer not to be coming to us humans as a potential conqueror. A successful planting of our Human Tree of Universality would suggest that the great debates about 'who we are as humans', and 'the purpose for our existence' - have yielded answers that most of humanity can now relate to. After this point, access to an external intelligence is not expected to materially change either of these two answers. We can now go forward, confident in our own identity, as we interact with the external intelligence as a single human race – as opposed to a multitude of warring factions.

These advanced being(s), having presumably waited for hundreds, if not thousands of years, would be anticipating the awakening of our new civilization with heightened expectations. Would our civilization come together on a normative Philosophy and Identity that would make it possible for it to quickly establish higher orders of encapsulation (Supra, Super Supra, etc.)? Or would we go on with our current emphasis at the bottom of the tree of Universality - and continue to focus on Level 3 and early Level 2 opportunities and threats? This second option would painfully and slowly bring about improvements through short term competitive behavior (continued biological and technical evolution) and might lead to a degree of Supra encapsulation in about a billion years (about how long it took to develop humans from the first Eukaryotic cells). With this approach, surviving that long itself would become an issue. For example, our lack of long term thinking could easily end up bestowing our hastily concocted machine progeny with such immense powers that we end up making the human race itself obsolete (as implied by Bill Joy, The Future doesn't need us, Wired Magazine, April 2000).

If we are lucky, and actually survive the billion or two years it would take for us to reach Supra level sophistication through evolutionary forces – we might still be unable to grapple with cosmic events like the burnout of the sun or a possible collision between the Andromeda and Milky Way Galaxies. There is a better way to improve our chances of survival and prosperity – and it all starts with the successful planting of the Tree of Universality!

The ideal trajectory of a highly successful civilization is that its normative ideas about how the universe can better itself spreads at the speed of light. The cone of causality for humans today hardly effects anything outside our home planet, and here too most of the effects seem to be detrimental to our continued existence on this planet. Yet, the potential exists for us to come up with a normative philosophy that would change the way that we view ourselves as a civilization, and feasibly even how the Universe views itself and its latent capabilities. This is the great promise that any resident higher intelligence in our vicinity might be looking for - quite possibly with parental pride. Yes, this is the spark of divinity that indicates that the great Human experiment

has Universe Class potential! As we have noted, our real journey as a Human Civilization begins with the planting of the Tree of Universality. In our case that would include a complementary role for both our human and machine offspring that builds on the best elements of both.

Complementary Fabric

As noted in Chapter 9, our human and machine progeny inheriting the human civilizational mantle can be woven together into a tight knit fabric based on their highly complementary capabilities. The fabric will need to be robust enough that individual human or machine threads could go off line and be replaced without losing the structural integrity of the overall fabric. From Figure 10.4 we know that there will be areas where both humans and MSCs can be complementary and ones where they will overlap. A large amount of trust will be required for each constituent to know that the pieces that they are not proficient at - are being covered in a dutiful and transparent manner by their complementary counterparts.

The dutiful piece is easy to understand – but the transparency part likely needs some explanation. Transparency is the process by which the thinking processes and critical supporting data are continuously exposed to others who might be asked to step in and fill our shoes if for some reason we became indisposed. Transparency does not mean information sharing only from human to human or MSC to MSC, as it is quite possible that the entity to pick up after us might be from the other side. A good degree of transparency also allows humans and MSCs to be able to rotate their duties and move on to newer tasks as opportunities arise. Let us take a simple example to describe the features of transparency and scalability in the context of a human family of four - supported by MSCs.

Example of Gasket Interfaces in a Complementary Fabric

Our example family unit has a Dad, a Mom, a Boy and a Girl. The family is supported by four MSCs – one each for entertainment, education, nutrition and health. The combined warp and the weft for this typical human family supported by MSCs might look as follows:

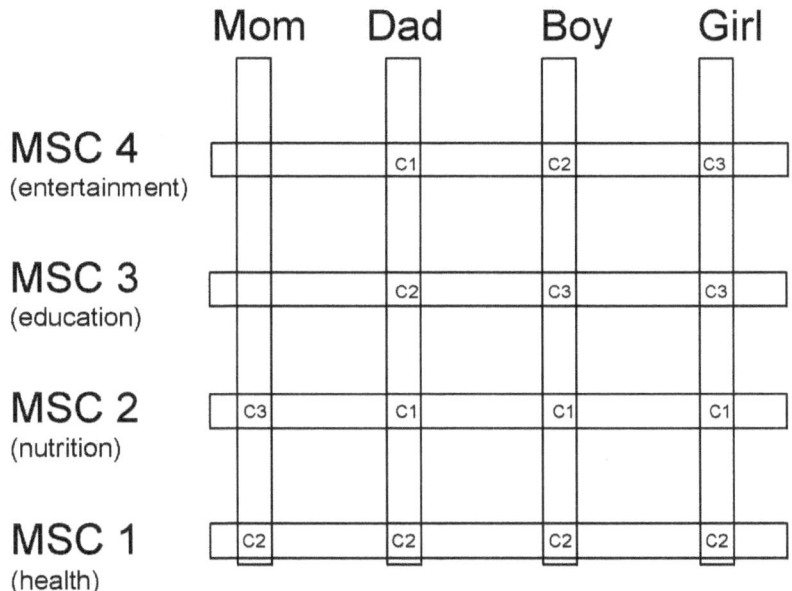

Figure 11.1: Social Fabric – example of a multi-MSC supported family unit

Let us summarize the four types of gaskets interface between humans and MSCs that we have defined from Chapter 7:

Category 1 (C1) – interface to outside world using our senses / muscles
Category 2 (C2) – senses/actuators displaced and acting at different location, time
Category 3 (C3) – human memory becomes machine resident
Category 4 (C4) – human thinking also become machine resident

In Figure 11.1 the overlaps of the human and MSC entities can support a range of gaskets, from a simple Category 1 interface, to a Category 4 with the human memories and thought processes both being resident within the MSC. In this example the highest we go is a Category 3 (C3) which depends on the MSC for the memories associated with the activities that are performed together. For example, in the education setting, it can be envisioned that the creative writing lesson that the boy is working on is totally resident within the education MSC - while the boy works on developing his thinking processes. The Dad, when monitoring the child's progress, needs a lesser degree of immersion, and is able to do so with a Category 2 gasket.

In the case of nutrition, it would be the Mom who interfaces with the nutrition MSC with a Category 3 gasket – retaining all the recipes as well as the family member's nutrition profiles within the MSC itself. The rest of the family can interface with this particular MSC at a more superficial level – like providing feedback on the size of their portions, and what condiments are preferred by the humans for a particular dish. Similarly the MSC responsible for health may not only have a C2 interface with the family, but support a C3 or even C4 interface with the family doctor who drops by electronically from time to time. In this type of a robust family fabric, the basic services are always well monitored and a high degree of safety and proficiency is maintained. Let us use this example to illustrate how transparency and scalability can also be maintained.

Let's say that Dad has to participate in a top secret project which requires that he be totally out of communication with his family for three weeks. The boy has been working hard at his creative writing, and the database is set up in such a way that the Dad has access to the lesson plans and the writing projects that the son is working on. In this example of transparency, when Dad is away the responsibility passes smoothly over to Mom to make sure that the son's education is proceeding as planned.

Now, let's say that while Dad is away Grandma comes for an extended stay. Grandma instructs her entertainment, health and nutrition MSCs to join forces with those of our example family. By

the time she arrives, the MSCs have worked together to integrate Grandma's specifics into all the nutrition, entertainment and health related activities. For the time that Grandma is visiting, she gets her afternoon ginseng tea exactly at 4:45PM – just like she used to at home. This is an example of scalability. The concept would still hold if Grandma and Mom decided to invite their whole extended family of 40+ individuals to come over for Thanksgiving.

What we have described here is a relatively simplistic example of how a tight knit human/MSC family unit might work. We can imagine that with the ability of the MSCs to span vast amounts of space and time, and hold memories and thoughts in essentially indestructible form - they would become the perfect complement to humans. We have also noted that the complexity of the MSC should be roughly at par with the human. This is to ensure that we can build a yogic symbiotic relationship where both the humans and the MSCs respect and value each other. As we have noted, the downside of waiting too long on building our first MSCs is that their intelligence would start off being so much higher than that of humans! The happy symbiotic relationship would now degenerate into one where the humans can't keep up with the MSCs intellectually – and the human civilization eventually becomes a historical footnote to a much greater machine driven civilization. This risk cannot be overstated!

The advantage of a well constructed man-machine fabric is easy to see in all aspects of the Mahashrama Cycle. A fabric that binds the two types of entities together will have the capability of responding to opportunities or threats within a fraction of a second – and yet have the capability to hold a focused attention on an area of interest for many years. It will have the ability to deal effectively from the details of Level 3 (Location Base) responsiveness, and also be able to scale the heights of Level 1 and Level 0 transformations that connect us to the core purpose of the Universe. Just as individual fibers can start and end, but the overall fabric can continue indefinitely; individual humans and MSCs can go in and out of a well constructed fabric – but the overall purpose of melding our collective vision into reality can go on unabated.

Convex and concave interfaces to machines

In the preceding Human – MSC interface discussion, the example fabric had the MSCs responsible to a large degree for basic household duties like nutrition, health, education and entertainment. Knowing how carefully humans need to take care of computers today, it may seem surprising that we would depend on MSCs for such basic necessities. To draw a parallel again from human biology, we can imagine the earliest computers as babies still in the womb – whose entire upkeep is dependent on the mother. The relationship might then be diagrammatically depicted as follows (Figure 11.2):

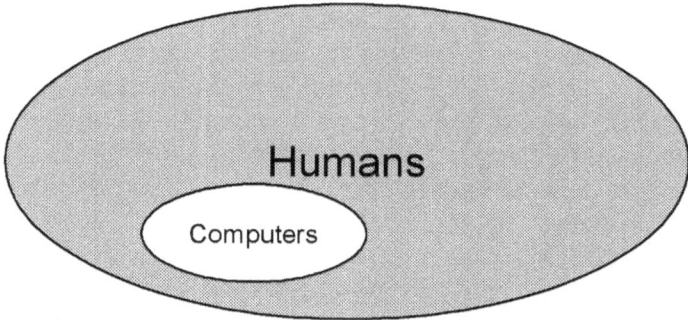

Figure 11.2: Computers in Bubbles – all upkeep from humans

Once computers started interfacing with the external world by themselves, say in applications like robotics and interplanetary spacecraft – the computer no longer remains encapsulated by humans, but still receive a large amount of its support and instructions from humans. This can be compared to a small child, with an attentive mother at close proximity to take care of things, and direct the child's safety and development. At this point, computers can be thought to have a convex interface – partially enclosed by humans – but not fully.

Over time as the computer grows up and takes on more of the responsibility aspects of the MSC – the interface begins to flatten out and then become quite symmetric. In other words, computers depend on humans for about as many things as humans depend on computers. This is depicted in Figure 11.4. If humans and

MSCs maintain their relative equivalence over time, they can continue thriving in a state of yogic unity. If, on the other hand, the machines end up taking full care of all humans, and receive very little in return, the man machine interface will have flexed over. This is depicted in Figure 11.5. It is as if humans were becoming the child (or the octogenarian) with their every need having to be met by the attentive machines around us.

The dependency shown in Figure 11.5, if it were offered to many humans today – might actually look quite attractive! After all, our lives would look like an extended childhood, and we would be free to go chase our every whim and fantasy – as we leave the hard work to the machines. Once the man machine interface flexes over, we might find humanity rapidly getting addicted on the 'free' labor and services provided by the machines. An entitlement mentality is easy to get into – essentially stalling the growth of the human individual through the stages of the Mahashrama cycle. This is a trap we must avoid for the very survival and vitality of the human race, not to mention our continued human participation in building a Universe Class civilization.

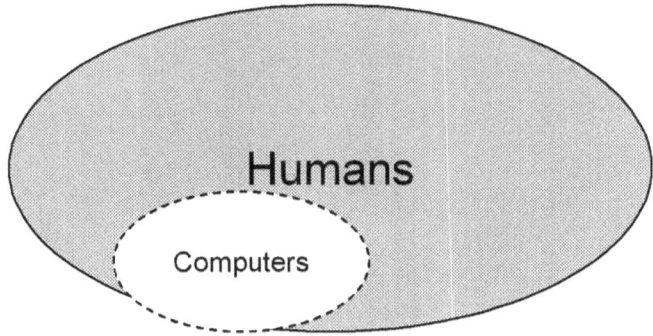

Figure 11.3: A computer "Child"

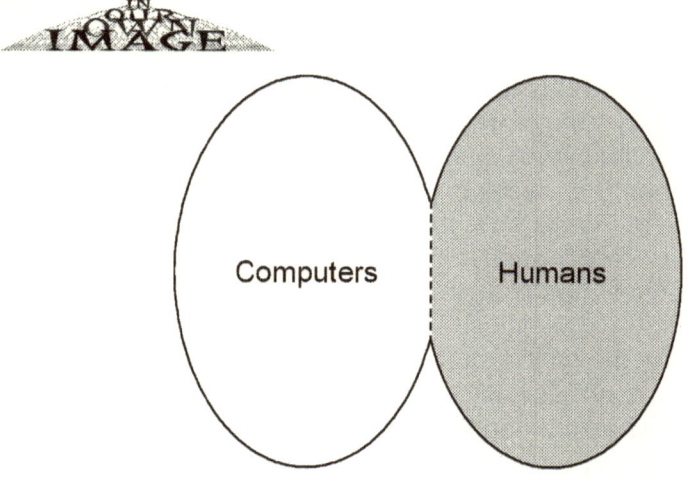

Figure 11.4: Human and MSC on equal footing

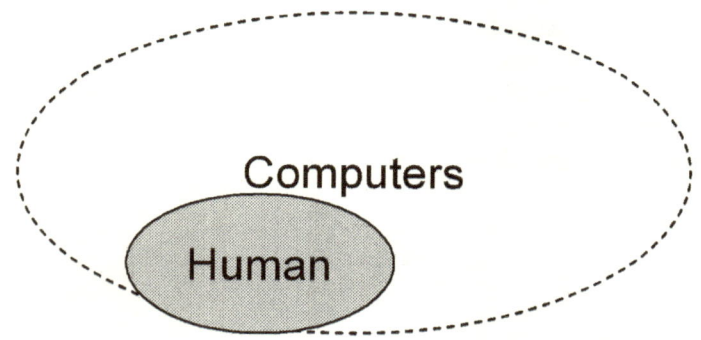

Figure 11.5: A Human "Child" supported by MSCs

Ideally, after establishing the human-MSC parity, we can build our way quickly towards the Sub-Supra and then the Supra entities. Human existence will now be similar to that of nucleated cells in the human body. Not all of our body is cells, as there are the various fluids - blood, lymphatic fluid, etc. - as well as non nucleated cells like the red blood corpuscles. We also have structural members like bone, nails and even hair which are largely made up of non-living matter. Similarly the humans (or groups of humans) that make their residence within a Supra entity would be like free cells or groups of cells organized together – supported within a largely machine based framework. Just like the vast majority of cells within our body interface only with

other cells and with the inter-cellular medium – most humans and MSCs inside a Supra will draw their interaction and sustenance from other humans and MSCs, or directly from their supra-cellular medium.

The level of organization of a Supra entity, of course, will put even the intricacies of the human body to shame – so we accept Figure 11.6 as a very crude representation of a superbly functional and intricately structured entity. With a total thinking capacity in the order of 10^{14} human beings, with localized response times in the order of microseconds and the ability to assemble a whole planet surface one atom at a time – the power of the Supra Entity would indeed be God-like. It is hard to imagine, but the same degree of exponential scaling from us to the Supra Entity can happen four more times before we reach the dimensions of the visible universe!

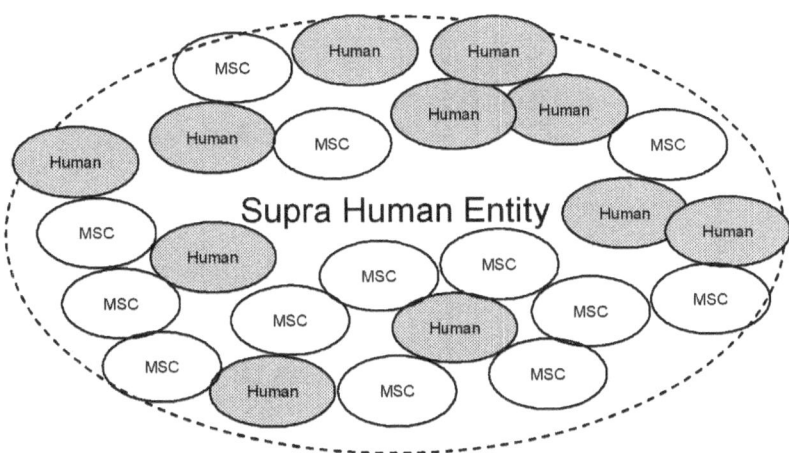

Figure 11.6: Supra Encapsulation

The binding of a support structure, vs. the binding of chains

On the flip side of an addictive dependence – we have the individuals who are not used to a social support structure, and are inclined to fend for themselves in the spirit of rugged individualism. These people may well be shocked at the idea of having to interface on a regular basis with other entities that support and

sustain them. The support can quickly grow into resentment - of lost independence, lost privacy, and the inability to do just what pleases us at a specific instant of time. If this sounds like the rebellious attitude of many a late brahmachari; and the cause for the breakup of many family units in early Grihastha – that is because the parallels are worth studying.

Like a spool of thread not yet woven into a fabric, the late brahmachari may not be ready to be tied down with responsibilities - or to be inducted into the predictable (but productive) patterns that define the life of the grihasthi. With the help of the MSCs, the onerous tasks that seem to consume our every waking moment as a grihasthi can be reduced, hence making it possible for us to have more time left over for the activities of our choice. We have already noted that the brahmachari is expected to receive more than they give from their MSC based support structure. With a person entering Grihastha, the MSC relations become much more of a two way street – with the MSC interface looking similar to Figure 11.4. Hence, the most desirable option after the L3 transformation is a balance of human/MSC capability – which avoids the dual slippery slopes of addictive dependence or of heightened individualism that refuses to be tied down to the predictable lifestyle of the grihasthi.

The value of such an extended social fabric is not lost on the maturing grihasthi as we struggle to bring up a family and provide a nest egg for our retirement years. If we have young children, we would not mind knowing where they are at all times, and that they are engaged in safe and meaningful activities. We would not mind if we did not need to dedicate our Saturday's to cleaning our house or restocking our refrigerator. And we would certainly not mind if the health of everybody in the family was monitored on a daily basis, and at the end of the day we were given an individualized portion of vitamins and minerals – only to make up for what we did not get naturally through our meals. Our individual medical records would remain our own, only to be shared with those medical professionals responsible for taking care of us – although the same machine based medical infrastructure might be servicing a city of millions.

Having established a synergistic coexistence strategy between our human and machine progeny, we can now turn our attention to building the essential infrastructure that is needed to scale the heights of civilizational accomplishment. Let us start with the core ideas and inventions that serve as milestones of progress for any civilization.

Gems that define a Civilization

As we look back through history, there are certain developments and ways of thinking that have given a significant boost in our quest to become the technologically sophisticated civilization that we are today. Some of these gems were the result of concentrated effort, but some others (like the discovery of penicillin) were amazingly serendipitous. In the days before digital timepieces semi-precious gems were used to form the low friction pivots and bearings within which wheels and rockers moved to and fro. In fact, the quality of a given timepiece was often indicated by the number of 'Gems' that went into them.

As we look forward to the growth of civilization, such epoch making discoveries like fire, the wheel, electricity, antibiotics and computers can be looked upon as the gems that forever shaped and channeled the direction and flow of our human existence. Similarly, in the world of ideas, the works of Copernicus, Galileo, Newton and Einstein have indelibly changed the core understanding of what we know of the physical realities of this universe.

In the realm of Philosophy and Spirituality, we have the works of the great Greek Philosophers (Socrates, Plato, Aristotle), as well as the ancient Indian wisdom that is captured (in somewhat cryptic form) in the Vedas and Upanishads. The philosophy of the great modern religions (esp. Buddhism, Christianity, Islam) are also gems in their own right.

The Gem-way

As we try to do in a few hundred years what would have taken a few billion years at the normal rate of evolutionary development, humanity will need to lay out a path studded with high caliber Gems that can then become the foundation of what it means to be supra-human. What we are talking about is the development of our next full level of encapsulation (the Supra), and comparable in complexity to our own evolution from the first eukaryotic cells. As we set out to design first the Machine Supra Cell (MSC) and then sub-Supra as intermediate steps towards that encapsulation – it will be important for us to look inside ourselves (i.e. how we work internally as motivated, open-minded, sentient beings); as well as look for external inspiration..

The remarkable thing about the Gem-way going forward is that this time the Gems will not come in linear fashion. In the evolution of humans from eukaryotes we can probably chart thousands of critical genetic developments – more or less in linear fashion that evolved us to the creature we are now. To understand the parallelism needed to construct the Gem-way, we need to go back a bit further in the evolutionary time machine – to the development of the eukaryotic cells themselves from their prokaryotic counterparts.

The Serial Endosymbiosis Theory (SET), which we have mentioned earlier, proposes that different types of single cell-walled creatures fused together to create the first eukaryotic cells with two cell walls – one around the nucleus, and one around the whole cell. Certain types of cell organelles (or subsystems) inside the cell like chloroplasts and mitochondria bear a striking resemblance to simpler prokaryotic cells in existence today – including carrying their own DNA, which is of a circular shape similar to most prokaryotes.

Translating this to generic terms, what SET implies is that the evolution of the subsystems that make up a eukaryotic cell evolved independently and were irreversibly fused together afterwards into a symbiotic relationship that lasts to this day. Taking a page from this evolutionary accident that likely shortened the normal evolutionary process by hundreds of millions, if not billions of years – we can lay out parallel projects that can independently contribute to the functionality of the MSC as described in Chapter 9. They can then be fused together into a well balanced and superbly functional entity. Of course, once the MSC has been designed and proven, we will have the even more monumental task of designing the internal structure and functioning of the Supra entity.

Coming back to the discussion of the Gem-way, what this means is that the pivots and bearings that define our future internal functionality will need to be spread out over a large range of human endeavor. It is possible that some of the pathways made possible by these Gems may not come to be utilized until the time of final assembly of the Supra entity or even a Super Supra (S2)! It is almost as if we humans embarked upon a crash course to identify and solve the technical problems that might one day be applied to the creation of the Supra being – however speculative the ideas might seem in the face of today's reality.

Architecting a Gem-way that is broad enough in scope to radically accelerate our ability to achieve our next level of existence within a few hundred years is not going to be easy. A set of humans of the long term Vanaprastha mindset would first need to think through the general architecture of what we are undertaking to build, and set out in general terms the Gems that will be needed to pave the path to the ultimate realization of this vision. This organization that we will entrust to undertake this tremendous effort – is the one we have called the Vanaprastha Network. This network will have to use their considerable influence to set in motion the research, development and skill building activities that would one day make possible a concrete realization of this vision. Let us take a look at some of the key action items that would lay the groundwork for parallel gem quality developments that could be later combined in the nature of the SET to yield a greater human civilization. Not surprisingly, the first objective is to set up the Vanaprastha Network itself.

Key Short-Medium Term Activities (next 1-100 years)

Job #1: Establishing the Vanaprastha Network

As we have noted in Chapter 5, it will be up to the extreme effort
of the vanaprasthi amongst us to mobilize the potential that is
latent within us humans. The first task is of segment balance
– and for this a way needs to be found for the vanaprasthi to reach
far and wide – and be able to enroll every individual who would
want to be reached into the overall purpose of existence. For
those individuals amongst us (primarily Brahmacharya) whose
needs outweigh their ability to contribute – the Vanaprastha
Network will need to come through as a highly individualized
support network. For those of us able to work hard to build a future
for our families, the Vanaprastha Network points out the most
rewarding activities, and also helps coordinate efforts for greater
achievement in a safe secure environment. For the vanaprasthi
amongst us, the Network exposes outlets for us to be of service,
and helps coordinate our efforts for maximum effectiveness. And
for the sanyasi – the Vanaprastha Network provides a conduit for
the crystallized gems of wisdom to be shared and made relevant
to our greater context. These are but a few of the attributes of
the Vanaprastha Network, which is capable of addressing the
smallest needs of the individual, yet scalable enough to grapple
with the good of all humanity – and if need be, to the cause of all
intelligence in this Universe!

It is one thing to say what attributes the Vanaprastha Network
should have, and yet another to go about building one. If some
of the characteristics of the Network, like scalability, look like the
characteristics of the Machine Supra Cells described in Chapter
9 – this is but one more reason why we need to start building the
essential infrastructure for the MSC almost immediately. As we
have noted earlier, the technologies that will make a MSC, and
ultimately a Supra or Sub-Supra entity possible – are likely to
be developed in parallel and then combined together in a grand
process of Serial Endosymbiosis. In developing the Vanaprastha
Network with its ability to act locally (maybe one atom at a time)

and still maintain a global perspective – we hope to seed several of the critical breakthroughs needed to build a truly cellular Supra infrastructure.

At no time will the effectiveness of a Vanaprastha Network be more tested than when there is a natural or man-made catastrophe just about to unleash itself. If we take the great Asian Tsunami of 2004 as an example – a future Vanaprastha Network could be out there monitoring the environmental factors – from crust movements to tidal waves, and simulating in advance where the danger zones may be located. Such monitoring on a millisecond basis is not the forte of humans, but would involve a machine based network capable of being at a constant state of alert. Simulations may not be very accurate today in predicting when such natural disasters will occur, but it is a much simpler task to figure out how energies already unleashed would be dissipated. In a hypothetical connected environment, every coastal village would receive an early warning to clear out, and those out at sea would get the message to stay in the 'deep' until the tidal wave had dissipated. In spite of the best effort at containment, not all devastation would be avoided, and here too the Vanaprastha Network would have a role to play, as we go about picking up the pieces. The Network then takes the form of an Individualized Service Organization – to provide the timely and appropriate help for people to get their lives back together. Waiting days for aid would be unnecessary – it would now be a matter of minutes or hours at most. The damage from hurricane Katrina and the accompanying levee breaches in New Orleans in 2005 are yet another example of the need for a network that can scale to the magnitude of the task at hand. We have noted in Chapter 4 several other segment balance related opportunities that could also be prime contribution outlets for a Vanaprastha Network.

In the path towards drawing humanity closer into a single Cultural Genome the establishment of the Vanaprastha Network is also the #1 enabler. There are reasons why such a network must be a secular entity, and it is even advisable for the Network to be developed independent of any specific governmental or political affiliation. The goal is to provide an umbrella organization for all humans – so care must be taken to avoid 'exclusionary' ties to any

one group. This is not to say that the concept cannot be piloted in a specific locality – just that the operating principles should be able to be scaled to all of humanity (and possibly beyond, to all intelligences). Let us look at Figure 11.7 to check out the other key tasks that the Vanaprastha Network will need to help direct - and how they relate to the Mahashrama Cycle.

Job #2: Constructing the Human Cultural Genome

Once the basic Vanaprastha Network is operational, and a connection has been established to all the major cultural constituents that make up humanity – the job of building the Human Cultural Genome can start. As we have seen in Chapter 7, technology is making it possible to preserve the essence of who we are - as individuals, communities or ethnicities – as embodied by the principle of MAP (Mutually Assured Preservation). Knowing who we are, what we believe, and what we stand for – are but the first steps towards building the human cultural genome.

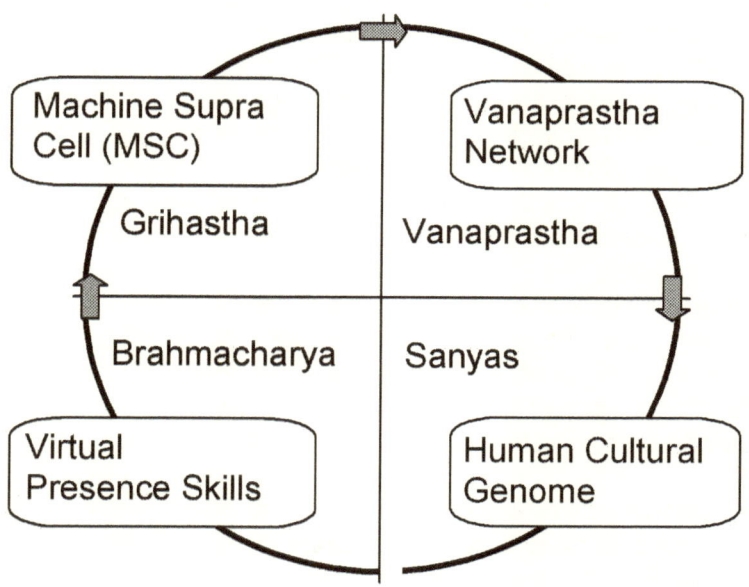

Figure 11.7: Key forward looking tasks – mapped on to the Mahashrama cycle

The human context today is extremely subjective – viewing the world as we do through the many lenses of language, culture, education and the opportunities we have been able to internalize. As an older generation passes away, so does most of the context around which they have lived their lives – especially in the scenario where there is no guardian intelligence to collect and stow away these potential gems for posterity. This human context is a terrible thing to waste – and more importantly we may be throwing away precious gems of wisdom that might mean progress or stagnation for our civilization at a critical juncture in the future. The time has come for us to start codifying what it means to be a unique individual or to have a distinct cultural perspective. We can then add them to the annals of recorded history as gems 'in the rough' that can be polished and beautified by a future generation. We have already noted that the technology to record vast amounts of information is already available and increasing exponentially. Recording multimedia descriptions of our lives is a start – but it is not enough.

The next step after the capturing of raw data is to put it in context. Let us take the example of a traditional wedding ceremony from the state of Bengal – in India. Just watching and listening to an accurate rendition of the entire festivities would not give an observer the meaning behind the customs and traditions, and the feelings that go through the minds of the bride and groom (who hardly know each other at this point) as they embark on a life together. To capture it all – not only the sensory piece, but also the memories and thinking processes that go with it require the use of a Category 4 Gasket as described in Chapter 7. We will look at the Virtual Presence skills required to replay this experience more in the next section.

We have also recognized (in Chapter 7) that the capability to simulate a real-time experience typically lags our ability to capture what we think is the most pertinent information regarding our unique perspective. Hence, capturing of the entire relevant context of an event, culture or viewpoint is more of an art form than an exact science. We will need to depend on the visualization skills of the active sanyasi to put themselves in a imaginary Category 4 gasket and find a way of capturing not only the facts, but also

the memories and the mindset (or thinking processes) of the key players. If this resembles how an accomplished novelist or film director goes about developing their characters and building up the context to tell a story – there is much learning to be drawn from such an analogy. Ancient historians dealt with the limitations of their limited medium when they told their story – resulting very often in a dull text based narrative. With the power of multimedia technology at their disposal, the stories that the modern sanyasi can tell can become a lot more 'alive' and immersive.

Capturing our individual and cultural viewpoints is a start – much like the primordial soup was a start for life as we see today. It is a very important start, because it puts the preservation of every individual's (or community's) unique perspective and identity in their own hands. Collating and distilling the entire human context into a clearly articulated Human Cultural Genome is the second half of Job #2. Here again, we will draw from the parallels of the Human (Biological) Genome to propose how the Human Cultural Genome should be set up.

If we look at human nuclear DNA, we find the Genes strung up in long DNA sequences that are encapsulated into chromosomes. There are 23 pairs of chromosomes – 22 of which have functionally equivalent pairings. One set, typically referred to as the sex chromosomes – are functionally equivalent in females (XX) but have a different pairing in males (XY). Leaving apart this unusual pairing structure of the sex chromosome in males, the two parallel and almost equivalent chromosomes is pretty much the norm. When we try to collapse the entire human experience down to the smallest number of internally consistent threads, the question needs to be asked: is a single perspective what we should shoot for – or could we live with two slightly different ones that bind the reality between them?

In paired chromosomes, it is fairly typical for long sequences of DNA to be identical between the two chromosomes. In fact, the genetic differences between humans is typically well less than 1% - and it is not much more than that when we jump species to say – a chimpanzee. But the sub 1% difference is sufficient to give us the various external physical attributes (like eye and hair color).

Perhaps more importantly, this diversity protects us from genetic disorders that arise from gene defects. This protection is there as long as we have one viable working gene out of the pair – much like a person with one working kidney can live a normal life. Let us see if this helps us answer how much cultural genetic diversity is actually desirable in the human cultural genome.

In Chapter 8 we have seen that the uniqueness or individuality actually rises for each level of encapsulation – from atoms to cells to creatures like us. Extrapolating how many levels (5+) we have left to go to get to a totally unique universal entity (Figure 8.5) – the conservative estimate would be to keep the internal differences within the human cultural genome in the 2-4% range. If it is much more than that – we have the situation where the internal inconsistencies would make it very difficult to make important decisions in a timely manner. The result could be a fractured internal population polarized by every topic that comes along. If we have too little (or no) diversity in the cultural genome – we risk recording errors becoming forever embedded in our genome. In the process we may also give up too much diversity for the sake of efficiency. From where we are today, codifying the collective human experience such that only 2-4% of the content needs a rebuttal – is itself a monumental task.

Coming to the rescue, in a big way, is the way we have defined Science to be predictable and repeatable. Using scientific principles, the nature of almost everything that we can isolate and study can be brought down to a single thesis. Over time even our internal functioning as humans will be well enough understood that a lot of the subjectivity can be removed. This applies to our understanding of our home planet earth, and most likely the other bodies around the solar system. Understanding that a lot of the functioning at atomic scales is dictated by quantum probabilities – the scientific principles would be able to be driven down into atomic and then sub-atomic domains.

On the other corner, defying efforts to write down a unified human story – is history. From our ancient past, and even now – what is recorded is but a very subjective account by a few, often biased observers. It is said that history is written by the victors in a conflict – which means that the minority (or losing) viewpoint

is often lost forever. Even after applying the principles of MAP – when it comes to history, it is very possible that the opposing viewpoints cannot be totally unified through Yogic Union. The hard part of creating the Human Cultural Genome will be to get all of the constituents with a vested interest in a certain version of history to cooperate in creating a description with the least amount of internal inconsistencies. After the sanyasis have given their best effort at distilling out a unified perspective – there may still be a parallel thread out there that describes with almost equal probability, another path that reality might have taken. In such a case, the alternate version would need to be captured in the Human Cultural Genome as well – and charged against the 2-4% genomic diversity budget. In case this budget becomes hugely overdrawn, it will be left to our future generations, with vastly powerful simulation capabilities, to determine which one(s) of the various alternative paths are likely the one(s) humanity actually traveled.

More important even than history in determining our identity and our cultural genome is our vision of our future and what we want to become – as individuals and as a civilization. The perils of having a highly factionalized historical identity can be mitigated by having a widely held unifying vision of what we want to be in the future. Even today, societies are termed 'progressive' or 'regressive' depending on whether the historical or visionary influences are stronger. The 'progressive' ones are striving to build themselves into the vision of a better future. The 'regressive' ones tend to emphasize more the past in defining who it is that they are, and how they channel their resources. To drive to a single human cultural genome – the importance of an overall vision of success for all humans cannot be over-emphasized. The visionary sanyasi will need to make it a priority to cross-correlate their visions, and build an integrative reality that appeals to all humans. We have already noted in Chapter 6 (Figure 6.7) and again in Chapter 10 (Figure 10.6) that even a subject that generates as much passions as religion can be put into an integrative context. The forward looking goal is to come up with a consistent yet scalable vision as to how our human civilization could become brilliantly successful - and to be able to contribute liberally to the core identity of our galaxy and potentially even our universe.

Job #3: Mastering Virtual Presence Skills

There's a Native American (Cheyenne) lesson about the way to truly understand someone else's perspective.

> *Do not judge your neighbor until you walk*
> *two moons in his moccasins.*
> *- Cheyenne saying*

For the brahmachari of the future, the best lessons that can be learned will be the ones where they can participate as a player and get the full perspective of what it means to be in somebody else's shoes - including sensory connections, memories, emotions and thinking processes. We have discussed the concept of the Category 1-4 gaskets in Chapter 7 – with each level leading to more and more immersive Virtual Presence experiences. After the skills have been sufficiently honed with the human contexts, the skillful brahmachari might be able to insert themselves into a variety of non-human contexts as well – say that of an animal, a spacecraft, or even an ecosystem. This skill is also critical in order to be able to create an effective human-MSC interface. A human using a gasket interface to an MSC theoretically can have access to all of the MSCs sensory systems, memory and thinking algorithms. Knowing the speed at which MSCs can process signals, it is unlikely that the human could access all of these sub-systems at full speed. What the composite entity would give up in raw speed, it would more than make up in other ways – as described in the warp and weft discussion earlier in this chapter. The ability to conjugate at higher gasket categories with an MSC is likely to be a rite of passage for the growing brahmachari – much like driving a car is for today's youth.

As noted earlier, our younger generation already has shown an affinity for Virtual Reality games – and Virtual Presence skills should be a natural extension. Today's games demand only a small percentage of the motor and sensory capability of the human – but a truly immersive Virtual Presence experience may actually exceed the total nerve capability of our brain. If we start

early enough during a child's development, it is possible that we could save some of the nerve capability that atrophies as we grow up. Beyond that, direct brain interfaces are a very distinct possibility. As noted earlier, early developments for a direct brain interface are already in place – as illustrated by the non-invasive skull cap with electrodes which could be used to move about a cursor on a computer screen.[2]

The utility of Jobs #1 and #2 are easy to understand for humans to have their impact at a cosmic level – the question may rightfully be asked as to why Item #3 even belongs in the same list. After all, playing video games is something children already do – and extending the Virtual Presence to educational and skill building sessions seems like a logical next step. Would it not be better if the neo Brahmacharya were to focus on those skills that were particularly suited to humans – and not bother too much about connecting with the machines as they go about securing the day to day tasks of operating our planet? What could be the downside of such a separation of duties?

Brahmacharya is the early development area not only for individuals but also for Civilizations. The human/machine fabric that would allow us to build the Supra entity - only works if there is a tight and systematic connection between humans and machines. At an individual level the rewards of having the ability to support a wide range of Virtual Presence gaskets are also quite obvious. A businessperson might be able to link in and visualize the whole crescendo of operations for his or her business – almost like closing our eyes and feeling the blood coursing through our veins. If something does not quite feel right – we can take proactive action to fix the irritants before they become real problems. An architect may be able to link in and visualize their creation well before it is constructed, and feel the effects on their structure as the winds rise, or a Richter 7 earthquake rumbles underneath. From a literary perspective, we can imagine the poetry that would be written by a connected human machine entity as they course the solar winds on the way to Mercury. This kind of conjugation is possible only if our human and machine progeny both develop the Virtual Presence skills that would allow the two to act as one. We might go even so far as to say that without a tight man/machine

interface and an expanding Virtual Presence skill set – humanity would never really be able to leave the confines of planet earth. This would seriously limit our ability to scale up our civilization to the level of our solar system (and beyond). Just like education today is a key determinant of future success, the young folks versed in Virtual Presence skills are likely to develop capabilities and skills far in excess of those who don't.

Job #4: MSC Development

> *Genius is one per cent inspiration and ninety-nine per cent perspiration. Accordingly, a 'genius' is often merely a talented person who has done all of his or her homework.*
> *- Thomas Edison*

The creative part - and the accompanying hard work of building the technologies that will become the underpinnings of our collective human future - will need to be the contribution of the Grihastha segment. The activities in this space are varied and complex, but the most significant achievement in the long path towards developing ourselves into a Supra entity – is the creation and education of the Machine Supra Cell (MSC). The capabilities laid out in Chapter 9 defining the key attributes of the MSC are but the tip of the iceberg. They draw inspiration from all segments of the Mahashrama Cycle, including some of the 'elevators' that move people between the segments. In the full scale development, this essential 'symmetry' between the segments should be maintained – and a conscious effort made to avoid the typical Grihastha trap – where all we develop is a super effective 'work-horse' that will obey our every command. The other key theme in the development of the MSC is their capability (along with our human progeny) to operate as the cellular divisions of the Supra Entity – modeled closely after how our own human cells interact to create an entity like us. We have also noted that many of these technologies would need to be developed in parallel. The parallel threads of technology would then need to be tied together in the MSC using the model of developmental acceleration exemplified by the Serial Endosymbiosis Theory (SET).

One key attribute that will need to be captured and integrated into MSCs is 'self organizing behavior'. Just as human embryonic cells grow and specialize to become the organs of the newborn child, the MSCs that come in contact with a new problem or opportunity must be able to grow and specialize, and thus behave like a single organism. The intricacies of self organizing behavior must therefore be built into the core human cultural genome that becomes the DNA of the MSC, and through the MSC of the Supra Entity. Without going into too many details, the learnings from how effective corporations manage their resources (that we have discussed in Chapter 2) can become an effective model for resource allocations amongst MSCs in the future.

Possible Objections to a Mahashrama based approach to human progress:

Objection #1: Religious - Machines don't have souls. We would lose our souls (essential identity) if we became overly dependent on machines. For a long time, the really orthodox have shunned new technology and learnings that might shake their faith in who they are as a people. With machines being so different from humans – it is understandable that the xenophobic instincts within us might rise to a fever pitch – and demand that we keep machines forever inferior and subservient to humans.

Logically, however, this is not a sustainable solution. Even if a small subset of humans were to find it rewarding to keep Moore's law going, and invent ever more powerful machines – sooner or later the resulting machines will catch up with us humans in thinking power and ability to control the environment. At the point where immensely powerful machines become cheap enough – they will be put to tasks that humans would rather avoid. Slowly but surely, humans will begin to specialize in the niches that they find rewarding – and leave the dirty work to the machines. The economics are just too overpowering – and only a few mystics and techno phobic communities (like the Amish) might be able to escape its brutal advance.

Since machines will make their advances anyway – the path of Yogic Universality suggests that we put an image of our souls (i.e. our human Cultural DNA) inside the machines – so that they can grow up as a part of us. Since we cannot win the war against the machines – the best we can do is make them our own.

Objection #2: Evolution has made us the masters of our domain, and Evolution will continue to bring about the developments to help humanity prosper. Evolution works through small sequential changes in our germ line, and the best future can be had by making small incremental changes that improve our current genetic and economic situation. There is no need for a radical jump in complexity as envisioned by the Supra entity.

Firstly, we clearly acknowledge the power of evolution as it relates to the development of cellular DNA. Slowly (but surely) we have evolved the features that make all life on earth more competitive and suited for their particular environment(s). We also recognize that it took more than a billion years for the first cells to evolve, and then another billion plus years for the eukaryotic cells from which we are made to evolve. Add another billion plus years – and we have humanity as it stands today. Evolution is powerful, but it is really slow!

With the growing emphasis on the right side of the Mahashrama cycle, it is debatable whether the mechanism or motivation even exists within us humans to weed out the unfit and the uncompetitive. There is also the ethical question of whether we want to live in a society that prunes the germ lines of those amongst us considered unfit or uncompetitive. Hence, the game has to change to where we play out all four quadrants of the Mahashrama Cycle – and acknowledge that human progress can happen through unselfish behavior and under the motive power of a fulfilling vision for humanity. Another way to look at it is – evolution is still at work, but the field has shifted from one of choosing cellular DNA – to one of choosing the fittest cultural DNA. The most potent cultural DNA can then become the foundation for the next stage of our collective human existence. Drawing this parallel, we also note that cultural and technological developments happen at a much faster rate than biological DNA

mutations – hence we can hope to cover in a few hundred years what would have taken billions of years using plain old genetic DNA.

Combining religion with evolution, it is interesting to note that there is a distinct possibility that the birth of our species and its evolution into something novel was intentional. Let us look back to the discussion on the descending astronomical factor – 108 – which we noted to be 3!. One of the implications we have drawn is that this may point to the presence of three concurrent intelligences in our neck of the solar system. One possibility we need to consider as to why we exist is that we are the third intelligence – along the lines of a father, a mother and a child! Yes, it is possible that Supra or even higher level intelligences will procreate sexually as well – and their way of creating a viable progeny is to mix their individual cultural DNA into a crucible or melting pot. Under this scenario very subtle background influences from our parents might exist, nudging humanity in one direction or the other. The central tenet that cannot be overstated is – when all is said and done our future is our own!

Objection #3: Humanity will never be united behind a unifying vision. If we look at the bewildering variety of languages, religions and cultures in existence today – is it possible that we can get a critical mass of humanity pulling in any one direction? Daunting as the task may seem – great religions and nations have been built in the past upon the same bedrock of volatile human interests that would lead us in a thousand different directions. To get started, especially on Job #1 – all we need are a few dedicated vanaprasthi to lay the foundation of the Vanaprastha Network. When it comes to building the Human Cultural Genome (Job #2) – we have the principles of Mutually Assured Preservation (MAP) – by which the individual perspectives can be indefinitely preserved. The governing process derived from the unifying principles of Yogic Universality, will be that whenever two perspectives meet – their Yogic Union will be greater than the sum of the parts. Similarly, the unifying vision itself will grow with the addition of each additional perspective. As in the functioning of any democracy – absolute consensus in all topics is not a requirement, and nor is it particularly desirable. The ability to forge

a critical mass to take concerted action towards desired objectives is already built into human organizations as simple as a family unit and as complex as an entire nation.

Objection #4: In the absence of total consensus – the inner workings of any complex society is subject to Internal Sabotage. Fact remains, even with an apparent consensus, the inner workings of complex machinery are subject to possible sabotage if there is something to be gained in the eyes of the perpetrator. In a tight knit society, early warnings can be detected and other avenues provided to address the perceived needs that would drive somebody to crime and sabotage. We have already seen that the Vanaprastha network would be invaluable here – with its individualized access to the entire community. We have also noted that in an advanced supra cellular medium there would be sensors everywhere, and it would be extremely difficult to do anything significant without a vigilant MSC getting wind of it. Given that the MSCs have the inner compass of the cultural genome built into them – they can now detect potentially dangerous behavior before these can actually cause anybody serious harm. For some humans, even constant monitoring may not be enough to keep them out of destructive behavior – and for these few some measure of containment may be necessary. This is not different from how highly secured areas are policed today, and the clearance and background checks that must accompany any person who is placed in charge of situations of great responsibility (or potential to do harm).

The fact that we have some redundancy and monitoring in the social fabric does not mean that we should not drive to a reasonable degree of consensus in our collective actions. This, of course, is the greatest strength of the Vanaprastha Network based on Yogic Empathy.

Objection #5: Humans have an innate need for a more powerful being. At some level we will resist taking our own destiny into our own hands – especially if doing so would confer upon ourselves God-like powers. The other concern with 'Playing God'

is that it might displease our actual God - the mere notion that humans might be impudent enough to confer such powers onto themselves.

The answer here too goes back to Chapter 6, Figure 6.8 and the image which we would hold out for our God. If we believe in a God from the Yogic Universality segment – we are ultimately part of our God – and hence this objection would be meaningless. It would imply, of course, that humans would carry themselves in a expansive and dignified manner as we went about our own glorious future – hoping one day for a degree of Yogic Unity with our own God. If our God is the Benevolent father figure from the top left segment, then we might want to behave like a child whose time has come to grow up and become an independent God-like entity. In fact, it is hard to imagine a God-like influence that would not want us to be the best, the most productive and successful that we could ever be.

Objection # 6: Intelligent Design – the claim that things in nature (including human beings themselves) are so finely tuned - that life as we know it could not possibly have arisen out a randomized set of events as suggested by the theory of evolution. This in turn calls for the unseen hand of the Creator – or God.

As we have discussed earlier (especially in Chapter 7) – there is a distinct chance that our life is a detailed reproduction intended to fill out the cone of causality. This is analogous to a controlled experiment designed to yield information and unique perspective within humanity that could be useful in filling out the existential gaps of an already evolved higher intelligence. Recognizing such a prospect, however, is different from requiring that this is the only way that life could have arisen!

Even if there was an Intelligent Designer, the effort seems to have been taken in the design process for our planet – painstakingly done, possibly one atom at a time – to recreate a naturally feasible time line. The idea of filling out a cone of causality does suggest that elements of the current timeline may be directly transplanted from some version of our current existence that actually happened elsewhere. By extrapolation, this would imply that if there is a

God or Creator – in which case we can infer that the intention of our creator is for us to find reality as we currently perceive it, and within these constraints build the most glorious future we possibly can. As far as the Prime Ascension goes – the glorious possibility that we, humans, along with our machine progeny can achieve the Supra level of organization unassisted by an external intelligence – it should not matter if we are truly alone, or we have a guardian intelligence watching over us but carefully choosing to stay in the background.

Objection #7: The human ability to make progress would break down when we come face to face with Immortality. Useful as the concept of transience is in giving humans a series of goals to be executed within a timeline – can we really grow up fast enough to grapple with immortality? Thus far we have only talked about immortality for our Machine based progeny – but the technology is not far away where even our biological brain may be repaired at the cellular scale and kept alive and vigorous for ever! If technology takes us there, will humans be able to deal with the new reality – or start jockeying for resources, not only to last a lifetime but essentially forever? What, then, happens to new population growth that we have coming on line – will we just end up crowding ourselves like sardines in a can?

Following the line of personal development embodied in the Mahashrama Cycle – getting closer to immortality will mean more individuals in the Vanaprastha and Sanyas stages in life. In the short to medium term – this is very positive. Over time, of course, the average transition age from Brahmacharya to Grihastha, and then on to Vanaprastha and then Sanyas will begin to get delayed; much as the childbearing age is getting pushed out in most of the developed world. Some amount of population growth is probably desirable – as long as the new entrants bring in the requisite skills to lead us on to greater individual and collective accomplishment.

The biggest risk of impending immortality is that humans will lose the immediacy of our limited time budget – and start playing the waiting game for conditions to be perfectly aligned before undertaking anything new. For this reason, a population made up entirely of mature individuals would not be the most productive or progressive. Hence, some amount of 'new blood' will need to

be transfused into any mature population to retain its vigor and forward looking ways. Yes, mankind can and will need to grapple with immortality – but with a bit of advanced planning this can to be turned into a largely positive transition.

Objection #8: The man/machine synergistic vision is only a temporary truce – sooner or later machines will not need us humans around any more. This outlook is presented by Bill Joy in his publication 'Why the future doesn't need us'. Interestingly enough, this is also part of the scenario described by the Unabomber in his 'manifesto'. The theme is – over time machines will take over the world, and THEY would have to decide, maybe along with a few human elite - what should be done with the masses of humanity who would be verging on obsolescence.

Given a reactive (we will see when we get there), or a repressive (never let the machines get too powerful) approach to the growing capabilities of machines, the doomsday scenario projected above is actually quite likely. Given a proactive symbiotic approach, along the lines of what we have covered in this book – we have the opportunity to fuse our human and machine identities together in such a way that it would be hard to demarcate where one starts and the other ends. We have already seen how humans and machines are very complementary in their skills and the way they go about approaching problems – and that a tightly knit man-machine fabric can have the best attributes of both.

The question remains as to whether it will be just the educated elite, or the masses of humanity that can take part in building a synergistic man-machine fabric. Again we have a planning challenge – how much of humanity can be reached by the message of hope and change in time enough to plan for the transitions ahead? Will certain cultures be more open than others in planning for the Virtual Presence skills that will allow the next generation of humans to meld more easily with their machine counterparts? Will a small splinter of humanity lead the charge towards God-like powers, and leave the rest of us in their dust?

Using the principle of Yogic Universality and a unified Human Cultural genome – the hope is that the vast majority of humans can join hands towards building a synergistic future. With the

direction coming from the Vanaprastha Network, the focus will not be on a power grab any more – it will shift instead to a firm belief that we can all grow and prosper together; that every perspective has value, and that with Yogic Unification we can build an indefinitely scalable and inclusive human civilization.

Along the ideas laid out by the Serial Endosymbiosis Theory (SET), if humans and machines can come up with a long term symbiotic relationship – much like the almost free living mitochondria within our cells – the cellular structure for our next level of encapsulation can be laid down to have both human and machine components. The question of the obsolescence of man is a relatively short term question in the cosmic scheme of things. Either we will come out of the next two hundred years with a clean symbiotic relationship with our machine progeny – or we will have to learn to live our lives as curiosities of history – much like we would treat the proverbial 'Missing-link' were he to suddenly appear in our midst! The stakes are indeed high, and get even higher for every day that we don't utilize planning and organizing ourselves for a synergistic future.

To prepare humanity to deal with the totality of the opportunities and threats at hand, it is no longer sufficient to measure our 'Intelligence' as our ability to solve 'Problems' and come up with discrete answers. Let us take a look at how Intelligence is commonly defined today, and the extensions we would need to be successful in the various stages of Mahashrama.

Higher Intelligence:

The ways we have traditionally measured Intelligence is primarily a measure of our intellectual fitness at about the stage we undergo a Level 3 transformation – i.e. during our late Brahmacharya when we have acquired all the skills to be successful in taking care of ourselves in this world. The skills tested – Verbal, Quantitative and Logical correspond roughly to the kinds of skills we build up through educational accomplishment. In fact the original trend proposed for IQ measurements was tied to age, and modeled a person's IQ (Intelligence Quotient) as rising gradually through childhood till age 16, and then staying flat through middle age

- till old age began to take its toll on mental sharpness. It is now acknowledged that some aspects of Intelligence continue to develop with age (esp. vocabulary); so a case could be made for why IQ would peak in the early 20's before beginning a long graceful decline. This corresponds roughly to when the brahmachari are at the peak of their skill building and are potentially undergoing their Level 3 transformation (Location Based Self Sufficiency).

A different measure of Intelligence that is proposed to be a better indicator of career success than IQ has been proposed – called Emotional Intelligence (or EQ). This is the kind of intelligence we would use in a socially interactive setting to bring out the best in ourselves, and also in others that we interact with. Part of cultivating our EQ has been the skill of being mindful of others, and the capability to align our goals with other individuals or groups.

> *EQ is not destiny—emotional intelligence is a different way of being smart. It includes knowing your feelings and using them to make good decisions; managing your feelings well; motivating yourself with zeal and persistence; maintaining hope in the face of frustration; exhibiting empathy and compassion; interacting smoothly; and managing your relationships effectively.*
> *- Daniel Goleman,*
> *Emotional Intelligence*

EQ is not as quantifiable as IQ – and is much harder to measure using standardized tests. Roughly, EQ can be translated to the degree of success at about the time of the Level 2 transformation – which occurs well into the Grihastha stage. Going forward, a high EQ person would find it much easier to be successful in the Vanaprastha stage with the strong emphasis on Community and Yogic Empathy.

Leading up from the Level 1 transformation through the Level 0 transformation – there is yet another aspect of Intelligence that comes into the picture – which we refer to as Spiritual Intelligence. Having undergone the successful Level 1 transformation, the concept of Yogic Universality for all forward looking intelligences

is now firmly implanted in the psyche of the successful Level 1 practitioner. At this stage we are working on knowing our core identity and the purpose that brings meaning to our individual and collective presence here on planet earth.

> *Spritual intelligence is our access to and use of meaning, vision and value in the way that we think and the decision that we make.*
>
> *- Danah Zohar*

Danah Zohar and Ian Marshall in their book 'Spiritual Intelligence: The Ultimate Intelligence' make a case for why Spiritual Intelligence is the intelligence that makes us whole, and gives us our integrity. The definition is core to how we have defined Sanyas - where we ask the fundamental questions with which we reframe the very context within which we exist – as individuals and as a civilization. It takes us right to the door step of a Level 0 transformation – and allows us to move our thoughts along transcendental planes that have the potential to reshape the definition of who we are and the essential purpose for our existence. The Spirituality Quotient (SQ) is even harder to quantify than EQ, but qualitatively can be linked to our 'Wisdom' and our ability to grapple with core existential questions. It also correlates with our ability to synthesize a greater integrative vision for all of humanity, as well as our ability to take a very complex reality and condense it down to a cultural genome that can become timeless.

Some of the external manifestations of a high Spiritual Quotient (SQ) are the appearance of serenity and contentment with being who we are, and the existence that we perceive around us. Interestingly, exposure to extraordinary 'beauty' also seems to bring out similar emotions – especially if we can resist the covetousness that often accompanies spying something that we really like. In fact, leaving apart the beauty of humans (which is often associated with suitability for mating) most beauty is associated with concepts like 'Simplicity', 'Perfection', 'Gem', 'Ideal', 'Sublime' or even 'Divine' – words that carry significant meaning within the Sanyas segment. Too often, as we go through life we do not acknowledge the tremendous degree of functionality and potential that lies within us, and the natural beauty and creative potential that lies around us. By the nature of our very

transitivity, the sanyasi has a heightened awareness that the beauty around us is often temporary. The sanyasi works not only to preserve beauty in its most elemental and functional form – but also to connect it to the overall identity and purpose of who we are as humans. This connection to our overall purpose ensures that when the Supra train leaves with the human cultural genome at its very heart – the beauty that we perceive can be shared by all intelligences for a very long time.

Having discussed the three main forms of Intelligence, and their relevance as we progress through the Mahashrama cycle, we can extrapolate that all three types of intelligence need to be cultivated and mastered as we seek to expand the human Mahashrama footprint. A civilization that exhibits a balance of IQ, EQ and SQ is qualitatively different from one that gets by based on IQ only. At various times, we have referred to a 'forward looking intelligence', which we will define now as having the balance of IQ, EQ, and SQ to enable a robust Mahashrama cycle for its citizens.

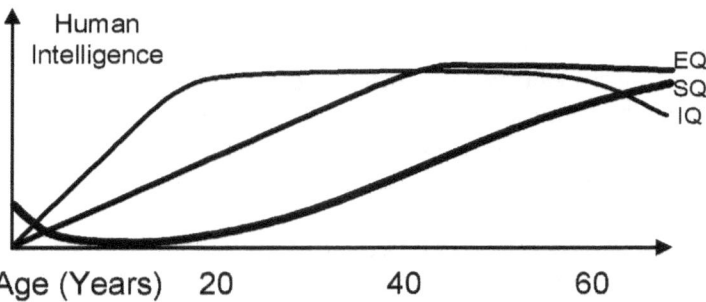

Figure 11.8: Projected IQ, EQ and SQ development for typical humans today

If we were to take a typical human going through the Mahashrama stages and plot out the growth of their IQ, EQ and SQ as they age, we would likely see a series of peaks for the various aspects of Intelligence (Figure 11.8). The IQ line rises fastest and flattens out by the time we are in the early 20's – at about the time of a successful Level 3 transformation. The drop off of IQ towards the very end of our natural lives is largely a function of the degradation of the brain itself with old age. The EQ line keeps rising until

we are well into our 40's – which roughly correlates to the age group of people in leadership positions worldwide. The SQ curve, however, appears to dip down as we approach our teenage years, and only begins to pick up as we come to the limits of our own IQ growth in the early 20s. Since we are yet to come up with a way to reliably measure SQ or even EQ, these observations are quite subjective, and could easily change if sufficient emphasis was put on the development of these intelligences as part of our general education. Yet, the outward manifestation of a high SQ is often quite easy to see – it is the spontaneous outpouring of joy that seems to gush out from an inner self brimming over with Ananda.

Moving from the individual level – to the Civilizational brain that we have called the Vanaprastha Network – yet another type of intelligence comes into the picture. This is the intelligence to apply the various intelligences already listed to the appropriate Mahashrama segments, and the ability to keep individuals moving between the segments as the opportunity arises. Let us term this type of intelligence to be the PQ or Philosophical Quotient of the Civilization. Needless to say, the current PQ of our Human Civilization is perilously close to zero, and can only be measured after we have put in place a Vanaprastha Network that can scale to reach all of humanity.

Ananda

As we started this Chapter, let us close this Chapter and this book by reiterating the message of Ananda. In our vision, we will be successful – and the outcomes will be enjoyable! The philosophy we can all stand behind is that no one who is willing to board this train will be left behind – that our plans will be scalable all the way up to the vision of an unifying intelligence across our entire Universe. The pain and suffering to which we owe our very existence today has been a necessary evil to get us to where we are as humans, through evolution and competition for survival. Going forward, a lot of the progress will be made through the use of Yogic principles – which are not as dissipative or painful.

The core message of life is about to change from 'Fear is the Key' (Alistair McLean, early 1970's) to 'Ananda is the Key!' Every system that motivates people towards action needs a reward mechanism. The reward mechanism is about to change from mere 'Survival' to 'Universal Citizenship' – with Ananda as its emotional embodiment. Being in Ananda is like breathing the air around us – we partake in this abundant resource as needed to sustain us – without any thought of consuming more of it than the other guy. The parallel to breathing goes farther. The first step towards learning meditation is often the use of Pranayama – or regulated breathing. One of the key goals of meditation, of course, is access to the well spring of Ananda. Hence, partaking in Ananda to our spiritual self should come as easily as breathing does to our material self.

We may be surprised by how close to the central tenet of existence 'Happiness' happens to be even from a perspective of nation building. From the perspective of national identity, Margaret Thatcher (ex British Prime Minister) once said:

> *Europe was created by history. America was created by philosophy.*
> *- Margaret Thatcher.*

An interesting bit of philosophy enshrined in the US Declaration of Independence is the inalienable rights to "Life, liberty and the pursuit of happiness" for its citizens. What Ananda promises is not just the pursuit of happiness but it's achievement! 'Pursuit of Happiness' works as an interim step until the citizens of the new millenium can learn to grapple with the 'Achievement of Happiness'. Ananda is our inalienable right; but we can only give it to ourselves! In recognizing Ananda as an inalienable Spiritual right, we also recognize that mastery of the Mahashrama Cycle gives us a working framework for reaching a high degree of spiritual accomplishment. Thus, Ananda gives us a metric for spiritual accomplishment - as individuals, as ethnicities, as nations, and ultimately as the technologically sophisticated human-machine super-entities that we are about to become.

Appendix 1
A closer look at Humanity's Security Needs

The time of this writing (2006) is a time for great cultural conflicts. As we have seen, the Safety and Security need is a very basic need for individual humans – and we can extrapolate that Safety and Security is a critical concern for us collectively as we try to chart a unifying course for humanity to transcend itself. Let us take the opportunity to take a deeper dive here, and see if we can come up with some answers that will enable the human Mahashrama engine to fire on all cylinders. To the responsible vanaprasthi answering the call of securing the future of humanity – the foremost question that has to be asked is – Security from what? Is it the security of the individual to take a walk in the park without fear of being mugged? Is it the security of the nation vying to secure its economic future by fighting to control vital natural resources? Or is it security from persecution of a minority suddenly made to feel unwelcome in their own country?

To put 'Security' in context let us discuss a story that is said to have happened in the early 2000s in a state in India widely known for its political corruption. A politically well connected individual works hard and saves up enough money to buy his first car. He goes to the local Maruti (local Indian car brand) dealership and buys himself a brand new red Maruti. On the way home he is accosted by thugs – who relieve him of his new car.

This individual feels extremely upset about the law and order situation and goes to the local political leader who was also a Minister in the state government. As he approaches, his attention is drawn to a Red Maruti – just like his own – that is parked at the Minister's residence. Curious, he walks next to it and confirms that this is indeed his car. Very upset, he walks in to the Minister's office and demands to know how his car came to be parked in the Minister's residence.

The Minister answers, "Oh, this is your car? Just give Rs 25,000 (about $600) to Mr, So-and-So and ask him to give you the keys." Before the owner of the car could really absorb the full meaning of this answer, came the parting piece of advise – "and if this happens again, don't forget to tell the perpetrators that you are my Man!"

This example may seem far fetched to some, but an alliance between the criminal elements and the ruling politicians is well known in many parts of the world. Politicians and bureaucrats in India (and most other countries) do not collect a huge salary. Ingrained in their thinking is the feudal aristocracy, where the lord collects the tributes from their subjects using whatever means available. The criminal element just happens to be the easiest way for those in power to enrich themselves.

A true vanaprasthi will realize up front that the responsibility (and accompanying status) they may get as a result of pursuing their duties – cannot be used to further either their personal genetic or economic imperative. This is the first level of 'Security' that the vanaprasthi must grant to all the people that he/she serves. This does not mean that a brahmachari or grihasthi needs to relinquish everything at the eve of starting their career of service. It just means keeping their personal and public lives separate and being able to switch between the corresponding value systems (Dharma) depending on whether they are on or off their official duties.

The Enemy Within:

The second entity that the general population will want security from is – society itself! Due to its very nature – having a large segment of society that is out pursuing their genetic and economic imperatives – conflicts are bound to arise. A well armed enemy in a foreign land is easier to deal with than an otherwise well adjusted person in our own country, and possibly in some position of responsibility, that suddenly snaps and becomes hell bent on destruction. In the fall of 2002, the Washington D.C. area was brought to its knees by an erstwhile member of the US armed forces and his teenage accomplice. Referred to as the Snipers,

the pair killed 10 people before being finally apprehended. In 2002, an Egyptian Airlines 747 was reportedly crashed by a co-pilot – killing 217 people – for no other apparent reason than that he just 'snapped'. Hardly a day goes by when we do not hear in the news of some otherwise normal family-person 'just snapping' and taking a gun to strike down their own family! The killing of the Nepalese Royal Family by the heir apparent (Prince Dipendra) in 2001 comes to mind as an example of how the best external security could not protect the Royals from an internal threat. The reason given is very close to the prince's genetic imperative – i.e. he was not allowed to marry the lady of his choice.

This is indeed a very intractable problem – and the first few steps have already been taken in many advanced societies to counter this. For those that may be tempted to take up crime as a last resort to feed their families – there is a social welfare system. For teenagers struggling to stay off of street gangs there are summer boot camps. For middle aged people trying to deal with an empty nest – there are counseling and psychological services. However, these are very preliminary steps and a lot more can be done to keep the well intentioned from 'snapping' in the first place. In places like India (and Nepal for that matter) the support infrastructure is much less developed. There the support would typically need to come from the extended family – and if the issue is at odds with the family itself (like in the case of the unlucky Prince) there is really no support available.

Cascading Failures

Taking the USA as an example, one of the big problems with the current system of safety nets is that – by the time the condition becomes apparent, things are already in a very bad state. And the safety net is often staffed in such a way that it often ends up making a bad situation even worse. In the early part of 2003 a young boy was found dead – almost in a mummified condition – in a plastic bin. The first question one typically asks in such a case is – where were the Mom and Dad? First, it appears that there is no responsible Dad – and that the mother was a single parent. So then, where was the Mom? The Mom was serving a 5 year term in jail for having abused another one of her children. So then

where was the State who had so graciously acted on behalf of the children by sending the Mom to jail? It turns out the administration had not had the resources to check on the Children for months, and had records that falsely indicated that a check had been done putting the children in good health – quite recently. The other two children were found starving and locked up in a basement by a cousin (with her own set of problems) who was supposed to be in charge.

This example shows a cascade of safety nets that ended up not fulfilling their function. In a traditional nuclear family, the first safety net is that there are TWO parents. If somehow one of them becomes indisposed, the other parent will typically pick up the load. With some human beings and in most of the biological kingdom – the male's genetic imperative is basically done as soon as a child is conceived. Biologically, the woman invests much more in giving birth to a child, and will usually stick it out as the single parent.

To accommodate for this huge asymmetry, society had created the institutions of marriage, fidelity, and monogamous relationships. These social constructs, and their enforcement, are all close to obsolete. The situation continues to get worse, not only in the developed nations, but also in the developing countries. The advent of AIDS has somewhat slowed the full extent of the Sexual Revolution that began with the widespread adoption of contraceptives in the 1960s. The deeply religious are fighting a losing battle as their 2000 year old social philosophy finds fewer and fewer takers. The current system of welfare (based on the number of children) and the 'Marriage penalty' sometimes perceived in our tax system only provide a further disincentive to dual parenthood.

The Vanaprastha Two-pronged approach

A two-pronged approach is necessary for the vanaprasthi to change the course of this emerging social disaster. There is a push and a pull necessary. The push is to build family and neighborhood solidarity – much along the lines of "It Takes a

Village to Raise a Child". This is the social fabric within which most of our needs can be met. But as we discovered from the story about the Prince of Nepal – not all needs can be met in this way!

The pull piece of the equation directly involves the vanaprasthi, and their ability to relate to individuals from all walks of life. This person to person contact can happen over large distances and can take many forms. In many ways it is similar to the relationship of the pupil to the Guru (or Spiritual Teacher) that worked so well in Ancient India. On a person to person scale we have referred to this skill as Yogic Empathy. Technology can help with both prongs – but there is a significant human element in all this that cannot be delegated to machines anytime soon.

Security from the Criminal Elements

Some societies do better than others in maintaining internal law and order, but the allure of quick gratification from illegal activities will continue to be there for the foreseeable future. Today we have the police force to maintain law and order, and the justice system to bring the criminals to justice. These institutions continue to remain invaluable – but the hope remains that over time the two prong Vanaprastha solution will help us keep these kinds of criminal activities to a bare minimum.

Security from a Usurping Majority

Towards the end of the big bull market in stocks in the late 1990's, it was estimated that Microsoft founder Bill Gates was worth as much as the bottom 40% of the US population combined. In a land without a strong rule of law, and constitutional provisions to safeguard personal wealth, the majority of the US population could have (hypothetically) decided to 'Nationalize' Bill Gates, and re-distribute his wealth amongst the needy. Far fetched as this idea sounds, just such a set of events ended up happening in Zimbabwe where a demagogic regime targeted the wealthy white minority and redistributed their wealth. The net effect in most cases is severely negative – as wealth passes from the hands of those who are most capable of producing more wealth (e.g. running efficient productive farms in Zimbabwe) to those that are ill

equipped to do so. The unique thing here is that some of the basic respect for property and other human liberties transcend the day to day law making by the majority! In the case of the US, much of these rights are enshrined in the Constitution – which has held up remarkably well for more than 200 years.

Security from Geopolitical Threats

We discussed in Chapter 3 the polarization that seems to be happening for most of humanity either along the Feudal-Theocracy axis or the Individualism-Democracy axis. Yes, 9/11/01 could happen again, but the solution space will need to be vanaprasthi to vanaprasthi – with the tools of Yogic Empathy playing a key role in coming to a common understanding. We talked in the last Chapter about the two steps of bridging a cultural divide and the doctrine of Mutually Assured Preservation. The same two steps are also necessary for Yogic Empathy to build at an individual level, a group level, an ethnic level or even a national level – it is just that the complexity of the task goes up significantly with each escalation.

Security from our own creations

It is expected that in another 20-40 years the average home computer will sport as much processing power as a human being, and then rapidly outstrip our capabilities in the years beyond. The calculations that lead us to such an estimate, as we have discussed in Chapter 9 – is based on a simple extrapolation using Moore's law and the exponential growth of machine complexity. It may take a bit longer for machines to have the degree of control over our environment that humans enjoy, but when they do, machines will also have the advantage of not having to sleep or take quite as many breaks as humans.

Much forethought must be given into the design of the future computing engine so that they are inculcated with a sense of identity and purpose that recognizes the Human Cultural DNA as the core identity of any future superhuman civilization. Following up on the Yogic unification concepts, we ought to be able to make sure that the core identity of what it means to be human is

preserved forever – even if it is an integral part of a much greater civilization. As proposed earlier, our best hope will be to emulate the symbiotic relationships that made possible the first eukaryotic (nucleated) cells – and chart a future where our machine and human progeny can exist and develop in Yogic Union.

Electronic / Information Security

Related to the technological explosion that we see happening all around us, more and more of our presence here on planet earth is now accessible from the electronic medium – also known as the internet. For example, almost all of the financial holdings that we have, from bank accounts to stock holdings – are held and managed electronically. These are increasingly subjected to crimes of identity theft. More and more people now have a personal web presence, like a personal web page or personal blogs. These are subject to the unique tribulations of the web like 'denial of service' attacks, or downright hacking and disfigurement. In addition we have the compilation of publicly available reports about us (e.g. our Credit Reports) that may or may not be factual. Going forward, as more and more of our daily existence becomes web based, it will be necessary for the vanaprasthi to introduce enough safeguards that the new electronic frontier does not turn into the 'Wild West' – the age of lawlessness and vigilante-ism that marked the early settlement of the American 'West'.

An alternate approach to Justice and Social Order

One thing we know about Security is that it goes hand in hand with enforcement, and a system for discouraging aberrant behavior. The system of Law and Justice today is based on rules that are mostly black and white – with relatively small amounts of discretion as to the relative perspective and motivations of the individual players. To a large extent, these laws are not much different from the days of the feudal aristocracy – where maintaining social order was of paramount concern. For example someone accused of stealing bread to feed their family would likely be found guilty of stealing. When the prescribed punishment is meted out - it might even include dismemberment. Someone

accused of stealing bread (or other essential items) three times even in places like the USA – can be put away in prison for tens of years under the 'Three Strikes' laws.

A system of justice based on the Mahashrama Cycle would be different. The skills of Yogic empathy would be employed to figure out if there are low order (Brahmacharya) needs that were left unaddressed – thus leading to deviant social behavior. Most crimes today are crimes of passion (an aberrant expression of the genetic imperative), or crimes directed at economic benefit (a deviant form of the economic imperative). Since these are both Grihastha needs, the first determination to be made is whether the person has made the transition out of Brahmacharya – i.e. do they have a degree of self sufficiency in taking care of themselves, with an appropriate level of understanding of the social and economic reality around them? The penal system now begins to look like remedial education, gating the graduation of an individual out of the Brahmacharya segment – and constraining their capability to partake in the genetic or economic imperative until a certain level of Mahashrama maturity can be established.

The four needs that have been determined as Brahmacharya needs can thus be considered in-alienable; and given to even the worst offenders. An argument can be made that the need for Individuality could be constrained as well for the hardcore offenders – especially if the crime may be derived from such individuality (e.g. racial hatred). The net result, then, would be that the responsible vanaprasthi would need to set up a two stage examination for when the offender could be graduated out of Brahmacharya – and the rehabilitation process could take considerably longer and likely be a lot more confrontational. Hence, taking away the need for individuality should be considered an action of last resort.

The test for entry (or reentry, after rehabilitation) into Grihastha then becomes a test of both the individual's self sufficiency as well as a balanced appreciation of the reality (Brahma) around them. This reality includes the needs of others – which we have seen is keenly dependent on the Mahashrama segment they are experiencing. In some way, negligently keeping a vanaprasthi away from being of social service or a sanyasi away from building a greater vision for humanity is also a misdeed – but this is a crime of opportunity. Social laws today do not have a good reading when it comes to crimes of opportunity in the Vanaprastha or Sanyas stages. In fact, some property rights can be downright detrimental when applied to the right side of the Mahashrama cycle. One example might be excessive use of Patents and Copyrights when it comes to core items of the human biological genome, or even the key parts of our cultural genome which will need to be adopted widely for them to become universally human.

We had started Chapter 3 (Mahashrama and Governance) with a quote from Aristotle.

> *I have gained this by philosophy: that I do without being commanded what others do only from fear of the law.*
>
> *- Aristotle*

For most rational humans, knowing and respecting the needs of others goes a long way towards avoiding behavior that jeopardizes the fulfillment of the needs outlined in Chapter 4. We have discussed Safety/Security at reasonable depth in this appendix, but the same logic can be applied to the fulfillment of the Physiological needs, or the need to get an Education. The basis of a civilized society now moves from just having 'law abiding' citizens, to having a population that is sensitive to each others needs and fulfillment processes. Yes, laws are still needed to enforce the gross transgressions – but by and large the population needs to be knowledgeable and respectful of the overall motivations of all individuals in the various stages of Mahashrama.

The other quote that we started off Chapter 3 with, again from Aristotle, is also very relevant:

> *All who have meditated on the art of governing mankind have been convinced that the fate of empires depends on the education of youth.*
>
> *- Aristotle*

In this case the education of youth corresponds to the graduation of the Brahmacharya segment with the right skills and outlook to be successful individually, as well as to foster a peaceful and progressive society collectively. A keen understanding of the Mahashrama Cycle and the needs that express themselves in the various segments are a necessary part of such an education. It is upon this foundation that the next stage of global human civilization would be built. As within the cells within our body – cooperation is much more the norm than conflict, and this too must be the norm within the greater civilization that we are to become. And yes, it will be up to the Vanaprastha segment to architect such a progressive educational system – thus leading to a future populace that governs itself only minimally by legislated lists of activities that we need to avoid and accompanying penalties if we don't. The emphasis now shifts to becoming more respectful and cooperative in the need fulfillments processes that drive our mutual progress through the Mahashrama Cycle.

Appendix 2
Tree of Universality – Mathematical Abstractions

Significance of 108 – Solar-Local Ratio

In Chapter 10 we discussed the Self Power Factorial (SPF) function as a mathematical way of decoding the Solar-Local ratio of 108. Built into our definition, we have seen that there are two terms (in italics) which could be any real number and yet preserve the following equation:

$108 = \textbf{\textit{0}}^0 x 1^{\textit{1}} x 2^2 x 3^3$

For Example: $108 = \textbf{\textit{100,000}}^0 x 1^{10,000} x 2^2 x 3^3 = \textbf{\textit{3}}^0 x 1^{15} x 2^2 x 3^3 = 0^0 x 1^1 x 2^2 x 3^3$

An explanation is probably necessary as to why these two terms behave the way they do. The two mathematical formulations that describe this behavior are:

$N^0 = 1$; where N is any real number (positive or negative),

and

$1^M = 1$; where M is any real number (positive or negative)

Acting together, the two independent variables N and M describe a two dimensional space, with N and M as the coordinates. Together they form a two dimensional abstraction layer (referred to as the veil of abstraction) that seems to separate out our physical world from the pinnacle of spirituality. In the olden times, when the Level 2 (Mahashrama) and Level 1 (Yogic Universality) was not built up enough to serve as a bridge to Divinity, it is understandable that an individual who did achieve Divinity (or Moksha) through spiritual development would need to relinquish all ties to the physical world.

One can imagine the 'N' and 'M' variable as two roulette wheels with a very large number of slots, both spinning rapidly. A spiritually developed being who came in through a specific

combination of these two variables, might look back after a while and see that the reality that he or she now perceives in the physical world, is not the same as the one they had left behind. Finding their way back would constitute a formidable task even for this spiritually developed being. Can this change when humanity transcends itself into a level of Divinity even in this physical world, through the evolution of the Supra or higher level of intelligence?

As a measure of Supra capability, we have previously noted that the task of building up a planet, one atom at a time – may be an interesting enough challenge for the Supra. Can a Supra, or a Super-Supra keep track of all the possible combinations in the veil of abstraction? Depending on how large the numbers M and N are, this ought to be another interesting challenge for these exalted entities. Since we will need to build a bridge for humanity to travel up through the veil of abstraction in sufficient enough numbers to bring about the 'Divinity' of our future existence – keeping track of the abstraction layer will become an important challenge for us in the not-so-distant future.

How big is the Veil of Abstraction for Humanity?

If the 'veil' is envisoned as a layer of abstraction, it is an interesting exercise to see how wide this abstraction layer must be in order to provide an individual connection for all of humanity. Let us consider how big the indefinite scaling would need to be in both Level 0 (Transcendence) and Level 1 (Yogic Universality) to uniquely cover every single human on earth. Until we get to 10 Billion people on earth, we can map each human individually on to the tree of Universality by using a unique combination of Level 0 and Level 1 coordinates – where each coordinate set is chosen to be an integer number in the range of 1-100,000. Yes, here again arises the surprising target number called the Laksham! This is because we can now have $10^5 \times 10^5$ combinations, which equals 10 Billion. The 'less than 10 Billion' era may not last for very long for humanity, but hopefully will hold till we can invent the MSC and lay the ground-work for the next level of human existence. If the veil were made out of 200 thread count cloth (i.e. 200 threads per inch), and each cross-over between the warp and the weft were a possible location to root a Tree of Universality, how big would this piece of cloth be to cover all of humanity? If it were a square piece of cloth, each side would be 500 inches, or about 42 feet (12.7m) wide. It is easy to see that without a superbly precise tracking mechanism, our individual existence can be hard to re-trace once we have placed some distance between us and the veil of abstraction.

Another interesting coincidence is that in ancient Indian mythology there are 330,000,000 Gods and Goddesses. This is close to 5% of the total population of about 6 billion human beings with whom we share this planet. If we cannot get all of humanity on board with the vision of universality, this is perhaps indicative of the minimum number we will need to proceed with, in order to lay the foundation for the future Supra Human Entity.

The Self Power function

The Self Power (SP) series, 1^1, 2^2, 3^3, etc. was introduced in Chapter 10, as a sidekick to the immensely powerful Self Power Factorial (SPF) function that we derived from the number 108. The SP series itself is no slouch, also outstripping the exponentiation function (10^N) as N gets large.

In a Yogic world, the SP function has some interesting properties. If we designate all yogic entities within a given population as being created out of two approximately equal halves (say physical and spiritual) – with each half able to support one to many associations with the members from the other half – the number of possible combinations follows the SP function. Let us take the example of N people each of whom has a house – i.e. for a total of N houses. Every Friday evening these N people may decide to all meet at any one house, or stay in their own houses – or any other combination of grouping amongst houses and people. Since each individual can choose which one of N houses they would like to go to – the total number of combinations becomes N^N – thus reflecting the SP series.

Now let us say that next week there is one more person (and their house) added to the mix. The total number of combinations now becomes $(N+1)^{N+1}$. If we started with just one person (and their

house) and keep building up week by week, the total number of combinations (including historical records) by which the N people get together in the Nth week follows the SPF function. This seems to suggest that if histories (i.e. time lapse views) are important, then we have to follow the more complete SPF function.

Another interesting thing about the SP function is that N need not be an integer – i.e. it can be any real number, positive or negative, which we will refer to as 'X'. If we plot out the SP function for X between minus 2 and plus 2, the graph looks as follows. In the negative part of the range, the SP function has an imaginary factor as well, which is indicated by the dashed line. If we were to envision the diagram in 3 dimensions (including one imaginary dimension), the wavy part would look like it is twisted around the X axis.

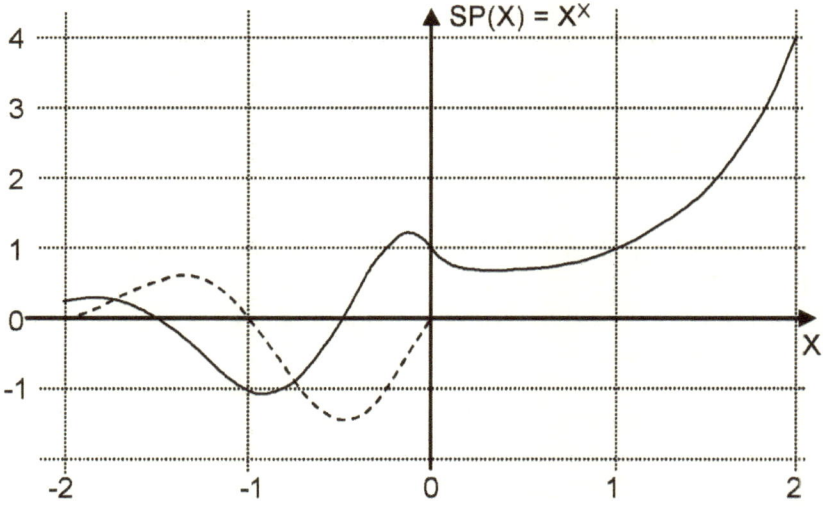

Figure A2.1: The SP function

Although we have taken the discussion of the SPF and the Tree of Universality down to Level 0, the characteristics of the SP function in the negative range suggests that the world is very different beyond Level 0. If Figure A2.1 is any indication, we are likely to encounter more and more wave like properties the further away we go from our current physical reality. This is not altogether surprising since our current physical reality is actually expressed

within a probability-wave based Quantum reality.

A key belief that hopefully comes across many times in this publication is worth noting again. Whether or not there are higher intelligences percolated through our section of the solar system – we must devise our own future as if we could count only on ourselves – i.e. we were Self Powered as a civilization. This does not mean that we cannot use any clues or inspiration that we might find in our surroundings that helps us to achieve a measure of Divinity. This belief is also not meant to disrespect any intelligence(s) that may be looking over us as our Guardian Angel(s). It is primarily meant to create a future civilization of great scalability and consistency that could potentially scale up to the magnitude of our entire Galaxy or even the whole Universe. The way to look at this approach is – if we did have a Superior Pre-existing intelligence here – what would we do of our own volition that would make our 'Parents' proud! After all, is it not the ambition of each generation that their next generation should have the potential to exceed them in their glory and wisdom? We might, in turn, be the proud parents of our progeny civilization that could aspire to reach for the stars!

The Peepul (Ashvattha) leaf, and the Bodhi Tree

Much significance is given to the Ashvattha tree in ancient Indian Philosophy, including in the Bhagavat Gita. The Buddha is said to have attained enlightenment under the Bodhi tree – which is another name for the Ashvattha. Today, this type of tree – the Ficus Religiosa, is commonly known as the Peepul tree. There is an unusual quality of the Peepul leaf that is worth noting. When the green leafy material is removed (either using chemicals, or through partial decomposition) it reveals a rich structure of veins and ribbing – which has often been used as a base for small paintings.

When we observe the Peepul leaf, especially with its ribbing structure exposed, it looks very similar to how we have pictured the Tree of Universality. Figure A2.2 shows the two of them side by side. The dotted lines reflecting the veil of abstraction has been removed for easier visualization of the Tree of Universality.

Figure A2.2: The Peepul leaf and the Tree of Universality

The first thing that jumps out at us is the similarity of the overall shape, with the fat bottom and the long pointed top. Looking deeper, we see that the sides are somewhat wavy as well, and the leaf is divided into approximately five sections before coming together at the top. This roughly parallels the three divisions of Level 3 (Location base) followed by the two divisions of Level 2 (Mahashrama), before uniting at the top (Level 1) and then tapering out at the very end (Level 0, or Transcendence). This visualization is admittedly approximate, but goes to show how a simple item from nature that the ancients associated with Divinity, bears a striking resemblance to the branching structure that is mathematically derived from a totally different source also associated with Divinity (i.e. the astronomical ratio – 108).

Notes

Chapter 1

1. The Mind and the Brain: Neuroplasticity and the Power of Mental Force – Chapter 3,
   by Jeffrey M. Schwartz, Sharon Begley; Harper Collins, 2003 (Paperback).

Chapter 2

1. The Innovator's Dilemma: When New Technologies Cause Great Firms to Fail
   by Clayton M Christensen; Harvard Business School Press, 1997.

Chapter 3

1. Web country databases: http://www.census.gov/ipc/www/idbsum.html; www.geohive.com
2. Aga Khan backgrounder, and speech: http://www.akdn.org/speeches/2005April7.html
3. The Theory Of Choice – A Critical Guide. Pg 126-127. Oxford(UK), article by Bruce Lyons (with Hargreaves Heap, Hollis, Sugden and Weale); Malden(Massachusetts-USA), Blackwell, 1992.

Chapter 4

1. The Maslow Business Reader – Part 1, by Abraham H. Maslow; Wiley, 2000
2. Self-determination theory and the facilitation of intrinsic motivation, social development, and well-being. American Psychologist, 68-78. by Ryan, R. M., & Deci, E. L. (2000)
3. Symbiotic Planet, Pg 31, by Lynn Margulis; Basic Books, 1998

Chapter 5

1. Gynecology services in rural USA: (Source: 2003 survey of 781 doctors in rural Florida).
2. Aging and Human Longevity – Part 1, Chapter 3, by Marie-Francoise Schultz-Aellen; Birkhauser, 1997.

Chapter 6

1. Empire: How Spain Became a World Power, 1492-1763 - Pg 107, by Henry Kamen; Harper Collins, 2003.
2. The Vedas, The Upanishads and the Bhagavat Gita, Pg 133, by Sri Chinmoy; Aum Publications, 1996.

Chapter 7

1. The Man Who Tasted Shapes; Section 2, Chapter 3, Pg 194, by Richard E. Cytowic; MIT Press, 1998.

Chapter 8

1. The Face in the Mirror: How We Know Who We Are – Chapter 2,by Julian Keenan, Gordon G. Gallup; Harper Collins, 2003
2. Redesigning HUMANS, Our Inevitable Genetic Future; Chapter 4, by Gregory Stock; Houghton Mifflin, 2002
3. The Arizona Republic, article by William Hermann Jan. 14, 2005

Chapter 9

1. Robot – Mere Machine to Transcend Mind, Chapter 3, Pg 54; Oxford University Press, 1999; – estimates human equivalent capability at 10^{14} calculations per second.
2. The Age of Spiritual Machines, When Computers Exceed Human Intelligence, Pg 103, by Ray Kurzweil; Viking, 1999; – 'conservatively high' estimate of human equivalent capability – 2×10^{16} calculations per second.

Chapter 10

1. Prokaryotes: The unseen majority; William B. Whitman*,
   David C. Coleman‡, and William J. Wiebe, University of
   Georgia
2. Symbiotic Planet, Pg 113 (Gaia, Chapter 8) by Lynn
   Margulis; Basic Books, 1998

Chapter 11

1. Redesigning Humans: Our Inevitable Genetic Future -
   Page 189
   by Gregory Stock; Houghton-Mifflin, 2002.
2. National Geographic, December 2004, Cap Harnesses
   Human Thought to Move PC Cursor

About the Author

Debashis Chowdhury was born in Shillong, India, and received his Bachelors degree in Electronics and Communications Engineering from the Indian Institute of Technology (IIT) at Kharagpur. Debashis moved to the USA to pursue a Masters degree in Electrical Engineering at the Pennsylvania State University. Upon finishing his course of study, he moved to Arizona to join Intel Corp in 1985. In the intervening years, Debashis has served in many capacities, including IC Design, Product Concept Development, Intel Capital (investments) and Marketing. Debashis received an MBA while working full-time at Intel. His technical work has also resulted in the award of an US patent (in wireless communications).

Debashis received an induction into Indian Spirituality at an early age. More recently, he was struck with the immediacy of the need to move our human civilization away from its dissipative ways – after being deeply moved by the ruins at Machu Pichu, the last bastion of the Inca civilization. Thus started the process of introspection and analysis, the results of which are documented in this book…

Debashis and his wife Sarbari are fortunate to have two children, Shilpika and Abhik. The family is one of avid travelers, and typically takes in several international destinations every year. They also maintain ties to classical Indian traditions, especially music, and typically host one or more visiting artists every year. The Chowdhury family makes their home in Chandler, Arizona.